DIGITAL
PHILOSOPHY

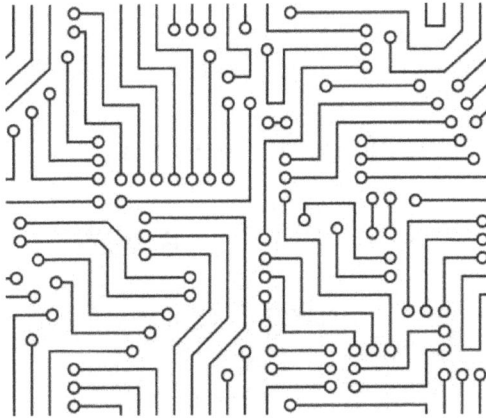

DIGITAL
PHILOSOPHY

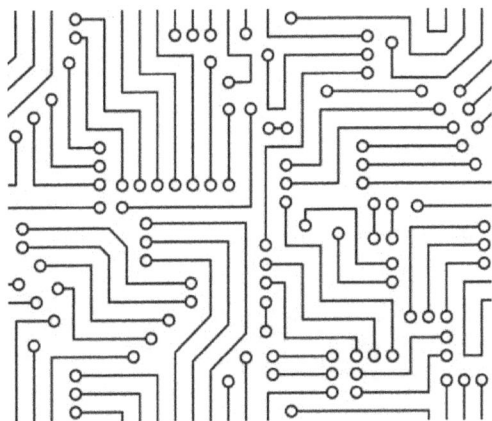

Mt. San Antonio College
Walnut, California

First Printing: 2014

ISBN: 978-1-56543-199-7

MSAC Philosophy Group
Mt. San Antonio College
1100 Walnut, California 91789 USA

Website: http://www.neuralsurfer.com

Imprint: *The Runnebohm Library Series*

Authors: *David Christopher Lane and Andrea Diem-Lune*

Dedication

To our two boys:

Shaun-Michael and Kelly-Joseph

Table of Contents

Acknowledgements

Andrea and I would like to express our deepest thanks to Frank Visser and *Integral World* for publishing a large number of our articles over the years and for encouraging us in exploring the frontiers of neuroscience, quantum physics, and evolutionary biology.

The MSAC Philosophy Group

MSAC Philosophy Group was founded at Mt. San Antonio College in Walnut, California in 1990. It was designed to present a variety of materials--from original books to essays to websites to forums to blogs to social networks to films--on science, religion, and philosophy. In 2008 with the advent of print on demand and cloud computing, the MSAC Philosophy Group decided to embark on an ambitious program of publishing a large series of books and magazines. Today there are well over 100 distinct magazine titles and 50 book titles. In addition, the entire MSAC database is now being put online via Amazon's Kindle, Barnes and Noble's Nook, Google's eBooks, and Apple's iBooks. A special mobile app called Neural Surfer Films is now available for Apple's iPhones and iPads, as well as one for Android operating systems on smart phones and tablets. *The Runnebohm Library* contains works on Einstein, Turing, Russell, Crick and other luminous thinkers. Some of the more popular titles include, *Darwin's DNA: A Brief Introduction to Evolutionary Philosophy* and *Global Positioning Intelligence: The Future of Digital Information*. Finally, *The Runnebohm Library* is in the process of producing a number of highly interactive texts that will include embedded video, games, and interactive feedback loops.

1 | *Global Positioning Intelligence*

I had an epiphany the other day (which I am sure I borrowed, consciously, from some other literary source) about how to interface a GPS device with what I will call Global Positioning Intelligence. What a GPS does remarkably well (I couldn't imagine not having one out at sea) is track one's changing position and momentum relative to one's changing geographical environment, whether in the ocean on the way to Catalina or in one's car on the way to San Juan Capistrano. More pointedly, a GPS is an information tool that is constantly being updated by electromagnetic waves bouncing off a satellite in space.

Now imagine this: Place a nano-sized GPS device, augmented with a nano- sized web browser constantly connected to the net but personalized with updates about the person wearing the GPS, so that not only is the person being tracked but the person is also housing tremendous amounts of personal information or data about who they are. Now imagine that you too have the same device implanted on you and with the sunglasses you are wearing (or the skin chip you have embedded) you have the ability to get an immediate reading (or display) of incoming blips on your data radar screen. So you walk down Main Street in Seal Beach to get a slice of Z Pizza and run into a stranger who accidentally bumps into you. Immediately (or even seconds or minutes or hours before?) you get a reading of personal data concerning that "ship" passing in the night replete with a personal profile. In sum, that person is a stranger no longer. You have his name, his occupation, his "my space" website--all available to you and within the privacy of your own head (or, more precisely, your own sunglasses or whatever augmented device one chooses).

That which used to be "hidden" is now externalized in bytes of information and wirelessly accessible without having to do anything except pay attention to whatever read-out is being generated at the time via your GPI. We are already there, actually. This isn't science fiction since Google Earth is already giving us a glimpse of what is here and all we have to do is bundle varying components together and we turn the private world public. I had intimations of this years ago when I tried to figure out how the world was going to get more psychic since it did indeed seem like that was the next stage our cultural evolution (even if we didn't want it).

MySpace, Facebook, blogging, and the rest are merely crude glimpses into what will be a truly psychic sphere, based not on some spooky mystical intuitions, but real nuts and bolts hardware grounded in quantum bits of data using nothing more than O's and 1's (represented perhaps by aligned/non-aligned atomic shards or even entangled photons) to discern our innermost secrets. Of course, this new GPI will usher in something unbelievably invasive and thus at first will only be self-selected by conscious narcissists or nerdy experimenters bent on pushing our digital envelope into the next "cool" app.

The iPhone is an example, albeit very rudimentary, of what is happening and what will happen. I remember just six or seven years ago writing in these pages about how upset a colleague of mine was because I had the audacity to post my diary online for all to see. Blogging hadn't caught on much then and I don't think he had any idea that the planet would become a blogsphere within a couple of years and that documenting one's moment to moment experiences would become an alternative lifestyle.

But scratch all that, the interesting chapter is when we all become walking or driving GPI's and such personal information pushes itself on us (versus the pulling we are doing now in the slow, type, mouse click, touch, or, scroll way, etc.). What does this portend? Well, to put it in literary terms it means that Aldous Huxley's *Brave New World* is going to mate with George Orwell's *1984* and, well, the GPI will be its most legitimate offspring. We are going to let the world in because we want to be known; we want to extend Warhol's 15 minutes of fame to 24/7 online availability. As Huxley rightly pointed out we are going to let us ourselves be controlled because we are going to entertain ourselves to death. And Big Brother didn't have to do it by getting us under house arrest. We will do it to ourselves, just like we are doing now, because the psychological need to be connected (even if it is a fictional connection) is greater than the need to remain privatized. We are a social animal par excellence and we are a psychic animal par excellence. Before we didn't have the technology to do what magicians faked in the past. Now we don't have to fake it, except in creating fictional Avataric lives (forget usernames, that was merely the first warning shot across the net divide), and we will truly become the actors and actresses that our virtual simulators (known as self-awareness) has already prepared us for. To analogize a bit here: the neuron is only as good as its synapses, axons, and dendrites. In other words, the neuron works because it is connected and given enough of those viable connections we have a brain that developed consciousness or self-awareness, which if it is anything is another name for a self-navigating system with the ability to virtually simulate that which is not there presently.

So where does this lead us to via the GPI? Get ready to go on stage, the world's stage, even if you don't want to. Google Earth has already put your private backyard online for close-ups for the world to see. Now it is only a question of time when we are going to put our mind on display for all to see. A Google Psyche is already birthing and we are its attendant midwives. While I realize that some of this sounds a bit far-fetched, but surprisingly I think it is understated.

Ironically, I realized long ago that the best way to keep your privacy is to act as if everything you do is in public. Only those who know you and love you and are intimate with you will realize how different you really are from that GPI profile you gave out so freely. Or, as Mikhail Naimy said long before we went digital: speech is at best an honest lie. To which I would add, the image of yourself that you project in a public space is the pathway to retaining your privacy. The Zen Koan of our freedoms: More disclosure, more privacy.

FIRST POSTSCRIPT

I wrote the previous article several years ago and the future has arrived much faster than I even imagined. As William Gibson famously pointed out, "The future has already arrived. It's just not evenly distributed." Just today, for instance, the *Los Angeles Times* (April 18, 2011) carried a front page story describing *How Smartphone apps use GPS to help singles connect in a Crowd*: "About a dozen smartphone apps enable people to connect at sports events, shopping malls and other public places, using the same location- based technology that tells you about a traffic jam ahead or whether there's a sale on jeans at a store you are walking past. Proponents say it's simply a modern answer to that age-old question: Where can I meet someone? 'The whole point is to facilitate real-life meetings,' said Christian Wiklund, founder and chief executive of Skout Inc., the San Francisco company that makes the dating app used by Bergmann and Riely. 'It takes a lot of courage to just walk up to someone. These are good ice-breakers.' Skout, which is the largest of the location-based dating apps, claims 5 million subscribers and says the average age of users is about 26. Grindr, aimed at the gay market, said it has 1.8 million subscribers. Most of these dating apps are free and work about the same way. People download an application and set up a profile that includes their photo, their interests and the type of person they are looking to meet. The apps are advertising- supported and offer added services for a fee. Subscribers to Are You Interested? can pay $1.99 a month to find out who has browsed their profiles. They can also send pictures of puppies or cupid's arrows (each cost 99 cents) to flirt with possible suitors. For digital wallflowers, Skout offers 'wink bombs' that send all selected nearby subscribers a pick-up line. Sample: 'Are you a parking ticket, because you have fine written all over you.'"

But these dating apps are merely the pilot waves of a future tsunami of GPI's. While meeting strangers for a hook-up is always an iffy proposition, meeting those with whom one is already aligned is a no-brainer (and, of course, the pun here was intentional). So it isn't a stretch to imagine that this year or next we will see dozens of apps using our GPS' in consciously intelligent and socially aligned ways so that we can (if we so desire, with the on/off option always looming large) gather in random places and in random moments with those who share our interests: from very specific religious affiliations to very specific culinary desires to very specific intellectual quests. The GPI we desire literally kills the Facebook of today. We want to create our own socially mediated universes and not merely live within and extend the dream of a Mark Zuckerberg. As Zadie Smith so brilliantly opined in The New York Review of Books, "I can't imagine life without files but I can just about imagine a time when Facebook will seem as comically obsolete as LiveJournal. In this sense, The Social Network is not a cruel portrait of any particular real-world person called 'Mark Zuckerberg.' It's a cruel portrait of us: 500 million sentient people entrapped in the recent careless thoughts of a Harvard sophomore." In order for us not get caught in the "locked-in" perversions of a Qwerty bounded shortcut for intelligent conversations on the Net, we will have to heed the words of Jaron Lanier's that "we are not a gadget." To have freedom of associations we have to be free from any singular association that imprisons us within its prefabrications of what is and isn't acceptable in our shared discourses. When we all become digitally telepathic and the universe at large becomes largely transparent then a new form of art will emerge where the personal narrative lives not in words or images but in self-generated worlds the likes of which we cannot envisage.

SECOND POSTSCRIPT

Google Glasses are the most talked about new gadget in the first part of 2013 and for good reason. They represent how intelligent devices are rapidly evolving to become embedded objects in our day-to-day lives. Already almost anybody with a smart phone (android or apple or otherwise) won't leave home without one and if they do somehow forget their intelligent devices they will invariably turn their cars around to go secure it.

This should give us a preview of what the future has in store for us, even if Luddites be damned. Everything and everyone is turning psychic and the hardware/software divide will quickly become lost in a mind meld that even Spock couldn't prefigure. Google glasses are merely an awkward transition from a partial smart augmentation (the iPhone or Android device is in our pocket, or our hands, or close up to our face) to a fully implanted one (from contacts to nano-technological seed implants).

What this portends is an unprecedented transparency of human cognition and connectivity. Ken Wilber has long argued that human evolution will move from the merely rational to the psychic, even though his prophetic trajectories were mistakenly of a wholly mystical kind. The psychic canopy of the future is not built on yogic visionaries, but on the nuts and bolts of hard core physics. The psychic template, even though it may seem to be imputing a spiritual realm, is algorithmically layered (level by level) upon electronic data streams prefigured in the laws of quantum mechanics.

And, yet, the larger question remains: are Homo sapiens ready for full and uncensored frontal lobe exposure? Perhaps Neil Postman's famous book *Amusing Ourselves to Death* should be retitled to "Exposing Ourselves to Death."

2 | *The Next Level is Psychic*

[*Diary Entry: 1999*] I had an unusual thought today. I rarely ever write anything down when I am driving, but as I was listening to my current book on tape (sidebar: books on tape are wondrously enjoyable) *A Biography of Carl Jung*, it struck me that Ken Wilber is right about the next stage in our conscious evolution. Wilber has postulated that we have undergone a series of holonic adaptations—from magical to mythic to rational--in our history. The next stage according to Wilber will be the "psychic" level. Ironically, what Wilber envisioned (a network of logical understanding that supervenes mere logic or rationality.... a sort of meta-thinking) is already obvious via the Internet and the Web. But it was Wilber's use of the word "psychic" that aroused my attention today. Usually when we say someone is psychic we think he can "read" our mind. The psychic has access to hitherto private information.

What struck me so forcefully today (after having my individual username held hostage and my email files hacked into) was how so much of our private life is turning public. What was hidden is now being exposed (by hacking, by hidden digital cameras, by innumerable information data banks, etc.). This morning I read that Microsoft's secret coding system had been sabotaged by a very sophisticated "worm" hacking system. Essentially these intelligent worms invade host environments and release "hidden" or encrypted programs that are invaluable to the company or person.

What is happening to our civilization? We are being wormed. What was internal is moving external. Everybody is turning psychic, but not in the paranormal sense of that term. The world is turning psychic in the following sense: We, as a people, are now getting access to the most arcane, secret corridors of companies, of religions, of organizations . . . of individuals. The individual's private psychic space is being outed. And, in this defined sense, we are all becoming psychic since we are getting access to that which was hitherto hidden. And, I must confess, it is a frightening transformation.

I don't think we are even aware of its devastating consequences. I don't think we are even aware of how intrusive such a development will be in the long run. Even in the short run, it is completely disorienting. And that very term [disorienting] is revelatory. Our "oriental" space (what we may term our internal, private closet) is being dissed. Our brains are being electrically exposed to the public arena. We are being hijacked and don't even know it. We as being are getting uploaded through a quick succession of digital confessions.

Our sins will be confessed not to the local priest, not to the local psychologist, not to our intimate friends, but to the audience. And that audience is ubiquitous. Why am I pontificating on this notion now? Because for the very first time in my life I realize that I no longer have any truly "private" space. The image is the reality. Why? Because the reality wants to be on T.V. It wants to be watched. Or, more precisely, the network wants you. And in so doing creates a unique entertainment event. Everyone on T.V. all the time. Everyone online available for downloading. Look at what happened to President Clinton. Every nook and cranny of his life got microscoped. And for what? The macro vision. . . The President as Daily Soap Opera Star. And we, I guarantee you, are Next. Everyone of us. Clinton's examined life will be, unquestionably, our examined life. And we are not psychologically ready for any of it.

Why? Because humans were evolved to have secrets, to have privacy, to have hidden chambers. Our triune brain is not a transparency but a surviving confusion. We are not even aware of our own secrets (enter Freud, enter Jung, enter Adler, enter hypnosis, enter dreams). But the world (as networked) will certainly have them in abundance. Not only Clinton's, or Carter's (remember his "lust" for other women), or Nixon's, but ours. You will be the T.V. You will be the Show.

And there is a guaranteed audience of one, if not millions. Are we psychologically ready to be mentally raped? I don't think so. The next hero is merely that person who keeps his psychic space private even under the tyranny of the herd mentality. Such a hero will hide in the most obvious of places. . . His Image. People will think that they know him better than anyone else because his honesty will reveal every last detail of his inner life..... And, yet, it was precisely in the arrangement of those images that he/she kept his privacy.

The Zen Koan of Privacy:

Be absolutely public. Only then, will you be left alone.

First I have a confession to make. I suffer from an incurable disease known as bibliomania. I love books. I love them too much. I roughly estimate that I spend at least 200 days out of the year going to bookstores or libraries scouring forgotten tomes to add to my ever-increasing collection. My obsessive love of books is arguably a genetic trait that I inherited directly from my father's literary DNA. I have had some wonderful adventures with books, ranging from the time that I was locked in Hayward's public library because I was sitting in an obscure bean chair and was so lost in reading Carl Sagan's *Dragons of Eden* that I lost all track of time and even, for a spell, of increasingly dimming lights, to receiving an OGSR grant from UCSD in 1987 so as to track down obscure texts in remote and dusty bookstores in Northern India. At first in my early teen years I couldn't afford to buy the books I really wanted, so I spent most of my time reading at the bookstore itself. The soon to be sold Bodhi Tree bookstore on Melrose Avenue in West Hollywood was my favorite haunt, especially in its earlier days when the crowds were negligible and a sweet little cat would roam the premises. My book tastes, however, when I was young were quite narrow, as I was enthralled with anything related to Indian philosophy. I think reading Paramahansa Yogananda's *Autobiography of a Yogi* at the ripe age of 11 is what influenced my focus.

Given my limited funds, it was difficult to satiate my lust for more expensive texts. One in particular caught my eye when I was 18. It was a leather bound edition of the *Guru Granth Sahib* in four volumes, replete with that incense smell that is the common lot of almost all books published in India. It was prominently featured in the now legendary glass case at Bodhi Tree. I had to have it. However, it was out of my price range (80 bucks), so I saved my earnings from working as a box boy at the now defunct Market Basket in Studio City until I could take home the object of my insatiable desire. I still have that same edition and it is prominently housed in my own glass case in my bedroom. Today I have something on the order of 10,000 books, scattered in three major resting places: my office at MSAC, my Huntington Harbour townhouse, and our larger home in La Quinta. My wife, Andrea, jests (though sometimes more seriously) that we bought the latter place just so I could expand my book collection.

I say all of this as a necessary prelude to the main thesis of this magazine: **The book is dead**. We are witnessing the funeral of the codex and the funeral pyre on which it is burning is going up in flames faster than we might at first imagine. I realize that most book lovers will immediately disagree with me, arguing for the uniqueness of the codex where one can easily navigate from front to back and in between in seconds. They will also rightly point out how a book has texture, aroma, and can be so easily portable. All of this I agree with and more. I love the feel of a finely printed book, with its gilded edges and acid free paper. Yet, something happened in the past few years that has changed my mind about the books future of the book and why we will see its present form evolve into something quite different. Indeed, the book is mutating into a completely new species, the likes of which would have been unimaginable in Gutenberg's day. It is the return of the illuminated manuscript, but this time with lights and sounds and interplay in a fashion that boggles the mind. I think we can date the day the book died fairly precisely. No, it wasn't the introduction of the Internet in 1969 (though clearly that was the first murder weapon). No, it wasn't the advent of the Web, as invented by Tim Berners-Lee, on one lonely night in December of 1990. No, it wasn't even the introduction of Sony's E-reader or Amazon's Kindle or Barnes and Noble's Nook.

No, the book died on January 27, 2010. This was when with much fanfare Steve Jobs announced that Apple was coming out with the iPad. At first the response to the touch tablet, which had been widely overhyped in the months leading up to it, was somewhat critical because it lacked features that some prognosticators predicted were fundamental (such as a front and back camera, the ability to display flash video, and multi-tasking). But within weeks after the iPad shipped there was an en masse conversion among a large number of skeptics once they actually got their hands (and this time the word literally does indeed literally apply) on it. I know for myself that within the first hour of getting it on day one I was completely converted. It should be noted that I was a first adopter to Sony's E-Reader, Amazon's Kindle, and Barnes and Noble's Nook.

While I liked the second generation Kindle (the Nook is too boggy for my tastes), I thought the grey screen and uniform print dulled the whole reading experience. Yes, reading it in the sun after surfing was a pleasure, but at home with the lights on I felt I was reading an old Etch a Sketch designed to have letters on it. Though I was duly impressed by the ease with which I could download books and even demonstrated how quickly the Kindle could do so to my varying classes at CSULB and MSAC (30 seconds or less in many cases), I was not wowed by the device and navigated back to books as my first choice for reading material. All that changed, however, when I got initiated into the iPad. It wasn't the iBook app (Kindle's app is far superior in many ways, since it allows one to access your Amazon library over a large array of platforms) or the beautiful way it put PDF documents onto the already built-in virtual bookshelf. No, it was the fact that I could read almost anything, anywhere, anytime while also accessing the web for news updates, Facebook updates, YouTube videos, Netflix streaming, online gaming, and whatever else strikes my interest or fancy at the time.

The iPad is, as other commentators have mentioned, the Swiss Army Knife of Information. And, as such, it has changed even the way I read a book. With a book I am in isolation, blissful solitude to be sure (just me and the writer in a digital stream), but we don't live in that world anymore. We live in a digital universe where vast Galaxies of data await our electronic mining simply by using our fingers. I understand too well the emerging argument from some psychologists that our attention span is getting warped by our increasing tendency to multitask and that we are suffering from digital overload. That is unquestionably true. I also understand too well the arguments first posed by the late Neil Postman in his severe critique of television and how we are "entertaining ourselves to death." I also quite appreciate that the iPad like its earlier mentor, iPhone, can do too many things and can (and does) distract us from the text in sight.

All of that is true, plus many more criticisms that one can find scuttling about on the web. But that is precisely the point. Language itself, especially as coded in books, is a virtual simulation of reality. It was never the real thing anyways. I don't see why one virtual simulation should hold sway and not evolve with our rapidly accelerating technologies. If we hold to these kinds of arguments, then one could (as others did back in the 15th century) justifiably lament the change from handwritten and idiosyncratic texts to mass-produced books with their uniform typography. We are not going back to sitting in caves and drawing animals on jagged walls, as breathtaking as they certainly are in Lascaux, France. And we are not going back to

11

the days of typewriters because we miss using liquid white out when we made misspellings. We are not going back—period— nor should we. Today one can access millions of books using one's iPad or a similar device without adding "weight" to one's library, since that library is in the "clouds", always accessible to anyone with an internet connection.

We are upending comedian Steven Wright's ironic joke about owning the world's largest seashell collection. It was so large, he said, that he had to keep it at all the beaches around the world. Instead of seashells we have books and we have so many of them now that we cannot store them in one place. Well, today my 10,000 physical books pale to the millions of books available to all of us in an instant. It is as if we went back in time and Cleopatra gave us the entire collection of the Alexandrian Library in Egypt and said, "It is all yours, do what you wish with it." But instead of accepting the amazing offer, we turn away and say "No thanks." Even this analogy comes up woefully short since we have about 50 times more books at our disposal right now at near the speed of light. All we have to do it reach out and "touch it." It is yours.

All of this became painfully too clear to me when a few months ago I walked into Barnes and Noble at Newport Fashion Island which is one of my usual habits (after visiting the always bountiful Newport Beach Library Bookstore) and realized that I didn't want to carry a hardback out of the store if I could download the same onto my iPad. While I still love buying old used books, I have found that I have little interest in adding to my collection with newer tomes from Borders and Barnes and Noble if I can get the same electronically. For instance, I have a hardback copy of the scientific treasure, Elements, and I have the same as an iPad application. The latter is so far superior in almost every way that I have relegated my hardcopy to the hidden shelves I have in my MSAC office. The book wants to be liberated. Information does indeed want to be free, but not in the monetary sense (I don't mind paying Apple or Amazon or Google their requisite fees). We live in a multimedia land and our books wish to reflect that as well. So instead of merely left to right text with three or four type fonts (Times Roman, Garamond, and Palatino Linotype being standard), we can have thousands of unique signature types augmented with interactive displays, streaming films, gaming tools, discussion boards, and the entire *Oxford English Dictionary* at our disposal with just a swipe of our hand.

Yes, in the digital multiverse, we can have it all. Once you open up the proverbial Pandora's Tablet, there is no turning off the binary light. The distance between the conveyer of information and the receiver is collapsing at an accelerating rate. As Nicholas Negroponte said as far back as 1995, "If it is not on the Net it doesn't exist." Just the other day I received in the mail a large reference book from Brill Publishers in Europe entitled *Handbook of Religion* and the Authority of Science. It is 924 pages and sells for over 300 dollars new. Even though I contributed close to a 50-page article to this tome, I would never have bought the work myself unless they sent it to me gratis since I was one of the contributing authors. Does it even make sense to charge that kind of money for a book these days when we can just as easily send it (without any manufacturing costs whatsoever) for free as a PDF file? I think not. Yet, why should we hold such information hostage to an antiquated system simply because we are bound by tradition? In addition, I can envision this same book to be infinitely more useful and applicable if it were an embedded text surrounded with all sorts of bridges and pathways to data that each of the essays touches upon. In my own article, I think the interested reader would not only benefit from seeing pictures and graphs and charts, but also by watching films that touch upon how science is advertised in certain new religions blossoming out of India. Imagine a book about 9/11 without pictures. Yes, I am not denying its usefulness, but I think the Chinese cliché that a picture is worth a thousand words has more merit to it than Neil Postman and other media critics may wish to admit. As I often remark to my students, Socrates allegedly wasn't a great fan of books since he felt that oral conversations allowed for a greater give and take and that books were too permanent and too certain for their own good. Interestingly, I think Socrates was right. The harder and more expensive it was to put words into some material form, the more difficult it became to correct mistakes. In fact, the very notion of a book is becoming so transformed in the process that new words are needed to describe precisely what it is.

Though I am hesitant to admit it, there is a tyranny to beautifully published books of earlier centuries. They tend to inherit an inflated status that is completely unnecessary and ultimately misleading. But today, with digital publishing (and print on demand) when there is a correction to be made we can do it with a simple keystroke. And with embedded texts on mobile devices like an iPad, hundreds, if not thousands, of interested readers can add to the book's import moment to moment with their own musings, creations, and even mash-ups. The result is the opposite of a static book. In fact, the very notion of a book is becoming so transformed in the process that new words are needed to describe precisely what it is. I am not sure of what neologism to invoke, but I do know that the majority of my reading today is on my iPad. But I have noticed a different kind of reading experience than with a traditional book. With my tablet I feel centered in the middle of a digital wonderland. Almost as if I was in the core of an infinitely turning and adapting wheel that has innumerable photonic spokes bringing in and out electronic messages in a kaleidoscope of colored variations. My imagination or my musings or more importantly my questions are now tethered to a global brain which can, given the ever growing electronic connections, provide me with continual adjustments or clarifications to whatever narrative I am presently engaging. Perhaps the book is finally morphing into its higher potential, something which was previewed in early manifestations but never fully realized because it took too long to actualize. The book was a precursor to hypertext with its notes, bibliography, and index. But how can one really follow a note if what was referenced resided in a library 30 miles away? Or, how can one justifiably cross-reference a source if such is missing from one's personal library? The book of the future, the book that we are now getting glimpses of on Apple's iPad, is hypertext fully realized. Text, all text, all images, all sounds, all films, all games, all interactivity, at the speed of light and in the palm of our hands and in our visual and auditory fields. The book needs to be liberated from its material corpus and fly, fly unencumbered, at the speed of electrons.

Yes, I still go several times a week and buy used books, especially if they are nicely bound or hard to find. Yes, I still love the feel and smell and binding of books. And, yes, I will still surround myself with their multifaceted charms. But the truth is that books are furniture and what we are discovering is that information in its quest for freedom doesn't want to be chained down to an elevated podium or locked behind cages in temperature controlled rooms as if readers were nocturnal data thieves. Lest we forget, books are a particular repository of information and tying such wisdom to one form and one form only (acting as if its early genealogy somehow grants it pride and privilege of place) is to neglect the real project at hand which is that we wish to gain knowledge and wisdom. To echo

Nietzsche we have in our digital adventures killed the book. But in so doing we are witnessing the resurrection of knowledge without pre-made boundaries. We are entering into the portal of Jorge Borge's famous "Library of Babel" where every book with every permutation is housed in "an indefinite and perhaps infinite number of hexagonal galleries." But unlike in Borges' fictional narrative we don't have to visit this physical labyrinth. Rather, the entire world's evolving library is sitting on our lap just waiting to be touched. It is a multiverse calling out for our total immersion.

4 | *Reverse Engineering the Brain*

It is one of those curious ironies that crop up from time to time when reading philosophers that the very mistake they accuse others of making is precisely the same one they make while penning their objections. Colin McGinn's recent review of Patricia S. Churchland's book, *Touching a Nerve: The Self As Brain* in the *New York Review of Books* ["Storm over the Brain", April 24, 2014] is a classic case in point. McGinn claims that because Churchland realizes that the brain is the basis of mental activity "your reason to stay a philosopher evaporates." But Churchland never says such a thing nor does she even imply it. Rather, she states the obvious: it is better to do philosophy with a deep understanding of neuroscience than without it.

Just as if you wish to understand why surfers enjoy riding certain waves over others (deep tubes versus mushy white water) a deep understanding of oceanography is better than none at all. McGinn erroneously alleges that Churchland's intertheoretic reductionism should cause her to "switch immediately to physics, since everything is ultimately made of elementary particles and depends on their activity." Yet, he conveniently forgets that Churchland argues for a *consilience* (to echo the famous title of Edward O. Wilson's 1998 book of the same title) of all academic disciplines, including math, physics, chemistry, biology, psychology, sociology, etc. It isn't one discipline versus another, but rather a deeper understanding of how each correlate and connects with each other.

Of course, McGinn unwittingly concedes Churchland's very thesis when he complains that she has garbled his "mysterian" position that consciousness cannot be truly understood because "there are special features" to it that make it intractable for humans to understand it. McGinn argues that his position is like "a blind man ignorant of the nature of color will never understand what color is (while remaining blind)." But isn't this precisely what Patricia Churchland was driving at when she opined, "you cannot understand the mind without understanding how the brain works." Perhaps McGinn should listen to his own example (about the blind not understanding the world of vision) and open his eyes to what Churchland really said in her book and not his own hurried caricatures of it.

Perhaps then he would better appreciate why she entitled her book, *Touching a Nerve,* and become less of a naysayer.

But then again I am not holding my breath for too long, since McGinn's Mysterianism is almost an article of faith amongst those who believe that consciousness, as such, is an unsolvable mystery from the very outset. McGinn, Nagel, and others in this camp seem to hold to a set of inviolate axioms, such as "our inability to imagine what it is like to be a bat is permanent, since our imagination is constrained by the type of mind we happen to have." Yet, is this apothegm as accurate as it self-assuredly posits? Do we really have an "inability" to imagine what it is like to be bat or a dolphin or Mickey Mouse? Who has scientifically predetermined the limits of our imagination and certified that it is permanent?

My problem with Mysterianism is that it is on-starting proposition and is akin to training a horse to win at the Del Mar Racetrack, but pulling out right before the competition begins by saying in effect it is impossible for her to accomplish anything so why bother at all. How does one know this unless one allows the horse to run in the first place? Likewise, how are we going to explore and perhaps solve the mystery of consciousness by declaring to scientists and philosophers at the outset that it is a no-win situation?

Yes, it is quite possible that consciousness will not be amenable, like unraveling the genetic code, to a proper scientific understanding, but it is really quite pointless to categorically claim its insolvability as a universal axiom. History is riddled with thinkers who declared that some mysteries would never in principle be explained but which were over time.

In addition, declaring that the subjective feel of awareness ("qualia") is a hard problem doesn't actually illuminate the task ahead, but can if invoked too often serve as a sort of meta warning signal which unwittingly thwarts would-be researchers from even tackling the issue.

Instead of worrying about absolutes or getting unnecessarily sidetracked science can make great progress by taking a practical route where it focuses on one particular area, such as what Crick and Koch have done with vision, or what others have done on olfaction or mirror neurons, etc.

I realize that too often scientists can jump the gun and attempt to prematurely explain away certain phenomena that have yet to be comprehensively explored (with NDE's being a telling case in point). But despite such hubris, we can learn from these more wild, and unproven, speculations because by their very insufficiencies we can get a clearer glimpse of what further work needs to be done. We should welcome such failures, since the progressive nature of science is literally paved with them. Yet, if we only succumb to Mysterianism, we won't venture anywhere except squirm about in our armchairs spewing out deductive witticisms about why such and such cannot be accomplished.

Consciousness studies needs adventurers, and even if these voyagers never do solve the great mystery of self-reflective awareness there will undoubtedly be wondrous islands of discovery that will help us in a host of other areas, as has happened in the past where one line of inquiry led to the most unexpected of discoveries which had nothing to do per se with the project's ultimate goal.

For this reason I am greatly encouraged by the push both in Europe and in North America for scientific teams to reverse engineer the brain since this will yield some startling information about how the brain works the way it does. Paul Allen, founder of the Allen Institute for Brain Science in Seattle, explains the practical focus in its initial stages:

"The basic idea is to instrument the brain at a very fine level of detail and measure all the parameters—from the diversity of cell types to the electrophysiology—in the mouse visual system, and from that reverse-engineer how it works. That's an amazing challenge, and no one's done it yet. We know a certain amount about neurons. You can do fMRI and watch parts of the brain light up. But what happens in the middle is poorly understood. We're hoping for breakthroughs in understanding cell communication and information flow in the visual system. That's what I placed a large bet on."

Of course, the empirical emphasis here doesn't preclude other approaches to studying awareness. To the contrary, it is merely one approach among many. For instance, this past week I had the wonderful opportunity of meeting with Dr. Sriramamurti, an emeritus professor of Sanskrit at Andhra University in India, and Mr. Preetham Tadiparthi, a distinguished software engineer, who took time off from their busy schedule to drive some 8 hours to meet with me at my office at Mt. San Antonio College. Both are keen followers of spirituality and have long been associated with Dayalbagh in Agra, India. They mentioned that Dr. Prem Saran Satsangi, the current spiritual leader of their group and a noted scientist in his own right, has been keenly interested in the scientific study of consciousness and has sponsored a number of Conferences in India and elsewhere encouraging a dialogue between Eastern and Western approaches. Dr. Prem Saran Satsangi just this past week gave a video talk to the 20th annual Toward a Science of Consciousness held at the University of Arizona, which was co-chaired by Stuart Hameroff and David Chalmers, and featured such diverse speakers as John Searle, Daniel Dennett, Roger Penrose, and Deepak Chopra.

I mention this because Dayalbagh has from its inception advocated an internal meditational discipline as part of its core philosophy and on the surface would seem to be resistant to entertaining purely empirical studies of self-reflective awareness. Yet, under Dr. Prem Saran Satsang's tenure, Dayalbagh has been at the forefront of advocating a variety of avenues for studying human consciousness, even when certain pathways contravene their own religious beliefs.

I know this from my own personal experience, as four years ago my wife Andrea and I were invited to give a plenary talk at SPIRCON, an international conference on the interface between spirituality and consciousness, that was held at the Dayalbagh Educational Institute in Agra, India. Although we couldn't personally attend, we created a 30-minute film presentation based on a 50-page paper we had written focusing on why there was a deep impasse between science and religion. Our thesis was a very simple one: The conflict between science and spirituality primarily stems from a linguistic confusion over what the term "matter" means and what it ultimately implies. Moreover, we argued that understanding the physical basis of consciousness was bound to be a fruitful one and should be strongly advocated instead of resisted by those most interested in mysticism. Our working pun (with its Koan-like implications) was "If we are just matter, what is the matter?"

In other words, one of the real pothers in consciousness studies is a linguistic one and as such has led to all sorts of unnecessary confusions in a field riddled with enough of them already.

I must say I was pleasantly surprised to learn from Professor Sriramamurti and Mr. Taiparthi that the spiritual leader of Dayabagh had in his most recent talk at TSC in Arizona actually cited (apparently without irony and with encouragement) a paper Andrea and I co-wrote entitled "Is Consciousness Physical?"

Professor Sriramamurti personally explained to me that according to certain schools of Indian philosophy that this physicalist approach has a long precedent and has been advocated by a number of rishis in the past. Indeed, there is a beautiful Sanskrit word for understanding the elemental basis of this universe. It is termed MulaPrakiti and is defined as "The root of nature [and] it is a closer definition of 'fundamental matter'; and is often defined as the essence of matter, that aspect of the Absolute which underlines all the objective aspects of Nature. While plain Prakriti encompasses classical earth element, i.e. solid matter, Mulaprakriti includes any and all classical elements, including any considered not discovered yet."

Professor Sriramamurti then went on about how though Purusha (Atman or soul) is mostly seen as distinct and apart from its material corpus, ultimately Prakriti and Purusha are one. They are different ends of the same stream.

Thus, the chief hurdle between the subjectivist and objectivist camps is to reorient precisely what we mean by matter and what it portends. Our language and our misuse of certain terminology have created a superfluous roadblock and as such have sidelined us from more productive dialogues with differing perspectives.

To invoke a sports analogy here, the goal of football is to keep moving the ball forward even if one cannot score a field goal or a touchdown. Likewise, the scientific quest for understanding consciousness should be a progressive one and getting mired in Mysterianism from the very start is a sure fire way to go nowhere at all.

Perhaps we can take a hint from the root word of matter, which originally derived from the Latin word "mater" which means "mother" or "origin." It is no accident that the greatest technological breakthroughs in human history have been when we tested our imaginary hypotheses in an empirical arena. This is the great testing ground for understanding whether consciousness is indeed an emergent property of complex matter or something that transcends the known laws of physics. If we persist in being solipsistic luddites our science will remain fastened, like our posteriors, to its cushioned seats and will not find anything new and unexpected.

This doesn't mean, of course, that the physical sciences will be wholly successful in this endeavor, but if it does spectacularly fail we will learn more in that endeavor than we would otherwise. It is one of those puzzling ironies in science that we can often achieve more by failing in testing our theories than in confirming them. But in order to do this, we must be willing to seriously pursue our chosen line of inquiry.

The Mysterian position, as posited by some quarters, stops such investigations before they even start, oftentimes with fallacious axioms that are themselves invented fictions. Reverse engineering the brain may in the end not solve the mystery of consciousness, but we won't know that until the project is completed. And, most tellingly, if the Blue Brain Project at the École Polytechnique Fédérale De Lausanne (in Switzerland) doesn't ultimately succeed in providing the smoking gun of self-reflective awareness, then that very absence will provide fodder for transforming our previous paradigms. Yet even if the big prize is not secured we will succeed in gaining hitherto unknown forms of knowledge, the likes of which would be unimaginable in centuries past. As the official Blue Brain Project explains its mission:

"The ultimate goal of the Blue Brain Project is to reverse engineer the mammalian brain. To achieve this goal the project has set itself four key objectives: Create a Brain Simulation Facility with the ability to build models of the healthy and diseased brain, at different scales, with different levels of detail in different species. Demonstrate the feasibility and value of this strategy by creating and validating a biologically detailed model of the neocortical column in the somatosensory cortex of young rats. Use this model to discover basic principles governing the structure and function of the brain. Exploit these principles to create larger more detailed brain models, and to develop strategies to model the complete human brain."

Both the Blue Brain Project and the Allen Institute for Brain Science are at their core practical endeavors and we are lucky that in their varied pursuits they have not succumbed too early to the Mysterian Temptation.

Frank Visser in a recent email communication with me raises several pertinent points concerning the study of consciousness:

"I can't help but feeling that no amount of string theory will throw any light on emotions, thoughts etc. Perhaps it's not so much that the mysterians declare consciousness to be a mystery, but only if you look for it in the wrong location. You can't find an inside thing such as consciousness on the outside, only outside correlates of the inside. What would you say?"

I agree that string theory, as a fundamental theory which gives rise to the four forces of the universe (and perhaps more)—gravity, electromagnetism, strong nuclear force and weak nuclear force—doesn't shed the clearest light on emotions or thoughts. But if we move up the materialist scaffolding project (Wilson's consilience hierarchy) there are emergent physical properties that do indeed illuminate how emotions and thoughts arise and how they function.

If we focus on the biochemistry of emotions, for instance, all sorts of interesting and fascinating information becomes available. As Jocelyn Selim explains, "Oxytocin, a peptide produced by the brain's limbic system, is released in both men and women during sexual climax as well as during birth and breast-feeding. Receptors located in the brain's dopamine reward system reinforce the good feelings that these activities bring. But oxytocin, like love, works in mysterious ways. In women, estrogen seems to facilitate the feel-good effects of oxytocin by moderating the release of adrenaline and other stress hormones. Testosterone makes men more susceptible to the fight-or-flight response and mitigates the stress-relieving effects of oxytocin. 'Although cause and effect is difficult to discern, oxytocin certainly facilitates social networks,' says neuroscientist Jaak Panksepp. 'And better social networks are associated with better overall health and increased longevity.'"[1]

If we follow Wilson's lead (and, with some necessary caveats, Wilber's holonic schema), each academic discipline is nested within an explanatory hierarchy, such that if one wishes to understand religion or theology, focusing on sociology first is elemental. Likewise, to properly contextualize how society operates, a deeper grounding in psychology is of great assistance. Therefore, when it comes to grasping how our minds work, we realize that concentrating on biology and chemistry gives us a tremendous advantage in properly framing why we think and behave as we do.

The fundamental differences between a dog and a human are not due to some astral influences, but due to varying neuroanatomies.

Of course, one has to be careful not to indulge in cheap reductionisms, since reducing our appreciation of Mozart to jiggling electrons is neither helpful nor instructive. If we pay close attention to how knowledge is contextualized within preceding systems then that which looks impossible to explain becomes more amenable to a scientific understanding. But this necessitates that we take great care and time in properly nestling informational streams within their natural (or emerging) order.

This is why Frank Visser is correct when he cautions us not to look for an explanation of consciousness in the "wrong location." I realize that for some theorists the wrong location is the empirical arena. However, as I have long argued, I think this is a premature conjecture at this stage and one which too often is intertwined with religious approbations.

We seem to have no difficulty in looking for physical causes behind our lost of sight, or hearing, or smelling, or touch, so I find it peculiarly odd that we resist trying to ground our understanding of consciousness within the physics of the world we find ourselves.

What really is the difference between an ant and a cat and a monkey? Is it really something "transcendent"? Or, is it in their physical composition? Practically speaking, if we focus on the physics of awareness we will be much more successful in our endeavors than opting for a spiritual first explanation.

The fundamental differences between a dog and a human are not due to some astral influences, but due to varying neuroanatomies.

The issue of Alzheimer's is a good case in point here. In such a state, it may be quite difficult for one suffering such dementia to explain to another (not under its spell) what it is like. However, just because we lack such a subjective (or "qualia") understanding of that state doesn't mean we simply stop looking for physical reasons for its onset. Indeed, it is exactly because we are confident that Alzheimer's is a degenerative brain disease that scientists have made significant progress in trying to locate how and why it arises with the ultimate hope of preventing its occurrence.

This doesn't mean that we discount the subjective experience that goes along with dementia. To the contrary, it is those very experiences that give us a benchmark on how our medical knowledge and treatment is progressing or failing.

The study of consciousness is best served by a multi-pronged approach, but focusing on the evolution of the brain and how it works first seems to be the most practical course at present. Perhaps the distinction between "outside" and "inside" will melt away once we get a firmer grasp of the mechanics of how self-reflective awareness arises from complex nervous systems. Perhaps not, but before we succumb to axiomatic Mysterianism, we should see what the Allen Institute for Brain Sciences and the Blue Brain Project discovers in their neuronal quest. Their quest doesn't preclude our own individual journeys, as each is not mutually exclusive. I see no inherent conflict in studying the brain objectively and in voyaging subjectively within, since each can better inform the other about what they find and dovetail their findings respectively.

The meditating Buddha as a neuroscientist isn't a contradiction in terms, but rather an enlightened proposition for where the future of consciousness studies is leading.

NOTES

[1] Johnson, S. *Emotions and the Brain,* **Discover Magazine,** May 01, 2003,

Jeffrey Kripal's recent article, "Visions of the Impossible: How 'fantastic' stories unlock the nature of consciousness" for the *Chronicle of Higher Education* has caused a minor stir amongst academics for its positive slant on paranormal experiences. Jerry Coyne writing for the *New Republic* has penned a scathing critique of Kripal's piece entitled, "Science is Being Bashed by Academics Who Should Know Better". Coyne summarizes Kripal's overall thesis in just two sentences: "People have had weird experiences, like dreaming in great detail about something happening before it actually does; and because these events can't be explained by science, the most likely explanation is that they are messages from some non-material realm beyond our ken. If you combine that with science's complete failure to understand consciousness, we must conclude that naturalism is not sufficient to understand the universe, and that our brains are receiving some sort of 'transhuman signals.'"

Since Coyne didn't tackle several of Kripal's more telling ancedotes, I thought it might be fruitful to do a point by point analysis of Kripal's claims and see how well his argument holds up to rational scrutiny.

I briefly corresponded with Jeffrey many years ago when I was using his book Kali's Child, a controversial psychoanalytic study of the famous Indian mystic of the 19th century, Sri Ramakrishna, in my World Religions courses. I got some heat for doing so from the Vedanta Society in America, since they felt that the book was misguided and wrong on key translations from Bengali to English. Indeed, I received a long letter from a Vedanta monk who presented a detailed rejoinder to Kripal's main thesis and because of this I wrote directly to Jeffrey to hear his opinion on the matter. Interestingly, Gerald Larson, Emeritus Professor of Indian cultures and civilizations at Indiana University (and formerly at U.C. Santa Barbara) castigated Kripal for being too reductionist in his analysis of Ramakrishna's personal biography. Pravrajika Vrajaprana, apparently concurring with Larson, wrote the following in the *Journal of Hindu-Christian Studies*:

"While the author describes himself as a 'digger' who uncovers hidden material, his methodology has been more like the rogue cop who plants evidence only to 'discover' it for the sake of manufacturing his case. Betraying his own bias, the author frequently uses misleading translations to prove his thesis. Examples abound, but a couple will demonstrate Kripal's technique: Ramakrishna goes into samadhi seeing an English boy who reminds him of Krishna. According to Kripal, Ramakrishna goes into samadhi seeing the boy 'thrice-bent in an erotic pose', and 'stunned by the cocked hips of the boy'. Yet neither of the two references cited by the author mentions 'cocked hips' or an erotic pose. It simply states that the boy was 'tribhanga' - bent in three places. On a different tack, 'maga' is translated as 'bitch' - thus verifying Ramakrishna's purported misogyny - when the word is merely a colloquialism· for 'woman'. Other distorted translations alternately transform Ramakrishna into a pederast or an onanist. . . . The above examples are only a sampling of the book's failings. Sadly, there are inaccuracies of one sort or another on a majority of the book's pages. Kripal's hypotheses are based upon innuendo, prejudicial translation, and cultural misjudgments. Obviously, this approach does little to advance religious and cross-cultural understanding, and that is a larger issue at stake. Can a reductionist approach such as this offer insight into a mystic's world? In the end, Kali's Child has value as a cautionary tale, for the reply it gives to this question is a resounding no."

I bring this up as a partial preface, since Jeffrey Kripal's core argument (in his recent article for the *Chronicle*) is strident in its anti-reductionist tone, which, of course, is ironic given his earlier reductionist tendencies in *Kali's Child*, though Kripal denies that was his intention since he claims to have taken a "nondual methodology."[1]

In any case, it is intriguing to see Kripal's mind at work when it comes to something he obviously feels very passionate about and to see how well his argument eventually holds up under tighter scrutiny. For clarity's sake, I have numbered my analyses:

1. MARK TWAIN'S PROPHETIC DREAM

Kripal begins his piece with a fascinating retelling of Mark Twain's alleged prophetic dream of his brother's untimely death, "In the morning, when I awoke I had been dreaming, and the dream was so vivid, so like reality, that it deceived me, and I thought it was real. In the dream I had seen Henry a corpse. He lay in a metallic burial case. He was dressed in a suit of my clothing, and on his breast lay a great bouquet of flowers, mainly white roses, with a red rose in the centre."

Twain was relieved to discover it was only a dream. However, shortly thereafter his dream vision turned out to be all too real. As Twain reveals, "When I came back and entered the dead-room Henry lay in that open case, and he was dressed in a suit of my clothing. He had borrowed it without my knowledge during our last sojourn in St. Louis; and I recognized instantly that my dream of several weeks before was here exactly reproduced, so far as these details went—and I think I missed one detail; but that one was immediately supplied, for just then an elderly lady entered the place with a large bouquet consisting mainly of white roses, and in the center of it was a red rose, and she laid it on his breast."

Kripal believes precognitive tales like Twain's are manifold and not to be so readily dismissed as mere coincidences, since he argues that "Most scholars have no idea what to do with such poignant, powerful stories, other than to dismiss them with lazy words like 'anecdote' or 'coincidence.'" In addition, Kripal claims that, "As with the heads of Hercules' Lernaean Hydra, however, with every story we so decapitate, three more, or three thousand more, appear. We are swimming in a sea of such stories, if only we could recognize our situation. We do not know how many such stories there might be, much less what they might mean. We do not know because we have never really tried to find out. Why, after all, would we study something that does not exist? 'Water?' the fish asks. 'What's water?'"

While on the surface it may seem as if Kripal's lament has some weight, the opposite is actually the case. Many researchers and scientists do indeed take paranormal claims seriously. In fact, they take them so seriously that they go the extra mile to find out whether or not they are indeed indicative of something truly "trans" personal. Moreover, as I have long argued, parapsychological claims, including Twain's own recollection, are better served by skeptics than believers, since the latter tend to take such stories at their face value and not dig deeper to find out mitigating facts and circumstances which may upend their spooky import.

If we merely accept Kripal's retelling of Twain's dream (suggesting that it is really beyond present-day science to explain it), we might bypass a more "Columbo" like investigation of the narrative which, as I will demonstrate, gives a richer background and context for us to better understand why

Twain may have had such a "death" dream of his beloved younger brother, Henry, in the first place. Of course, this doesn't a priori dismiss the possibility of the paranormal, but it does ground the entire story with a chest of illuminating facts.

First, it is important to remember most human beings dream nightly and a large number of these have something to do with future events. The real question that must be asked is what events during our day-to-day lives are engendering or prompting specific dreams about our close kin dying. In other words, were there any circumstances in Mark Twain's relationship with his brother that would have prompted him to dream about his death? Or, is such a dream something that was completely unwarranted?

It turns out that Mark Twain, who was deeply fond of his younger brother, did in fact worry about Henry's well being for weeks and months before his tragic accident. Twain introduced his brother to a more adventurous life and they took 6 boat trips together on the Pennsylvania where Ron Powers recounts that Henry "labored at the bottom of the boat's labor chain." It was not a pleasant duty at all, since the pilot, William Brown was a "seething and abusive man." Life upon the ship was so bad, in fact, that Sam and Henry got into a heated fistfight with the Brown, which almost caused the 486-ton steamship to run asunder at nearly 15 miles per hour since no one was monitoring the steering.

Mark Twain and his brother even talked about steamboat disasters and what one should do in such a situation, agreeing that they would stick with the ship as the best route to safety. As Twain himself recalls,

"The subject of that chat was, mainly, one which I think we had not exploited before—steamboat disasters. . . . Henry remembered this [our plan of action] afterward, when, the disaster came, and acted accordingly."

Looking over Mark Twain recollections prior to Henry's accident, and given the volatile nature of steamboats in those early days, is it any wonder that one might dream of a tragic mishap? I think not, particularly knowing the troublesome nature on board the Pennsylvania, long before Henry's tragic accident.

As a parent myself of two young boys (and keeping in mind that Mark Twain felt a paternal responsibility to his younger brother, Henry, who he called "the flower of the family"), I often virtually simulate all sorts of potentially dangerous situations concerning my children so as to be better prepared in how to respond if such an unfortunate occasion may arise. I have even had a series of very vivid dreams about them, both positive and negative. Therefore, I am not wonderstruck by Mark Twain's dream, since a closer analysis of its specifics doesn't turn up something that is out of the realm of reasonable possibility—whether it be Henry being fitted in his brother's own suit (note: Mark Twain had money; Henry did not) or laying in a metallic burial case, since such coffins were increasingly popular during that time period.

I can well understand that one may wish to believe that Mark Twain's apparent precognition is indicative of a supernatural vision. But given that there innumerable prophetic dreams which never come true it seems much more likely (especially given the background which preceded his dream) that Twain was simulating a very real possibility and that his dream was an archetypal representation of what many of us fear may happen to those we wish to protect from harm's way.[2]

2. A DREAM FROM "BEYOND KNOWING"

Jeffrey Kripal follows Mark Twain's dream with a story from the book, Beyond Knowing, recounting a remarkable tale of a wife who dreams "her husband standing next to her bed, apologizing and explaining that he had been in a car accident, and that his car was in a ditch where it could not be seen from the road. She awoke immediately, at 4:20, and called the police to tell them that her husband had been in a car accident not far from their home, and that his car was in a ravine that could not be seen from the road. They recovered the body 20 minutes later."

Kripal is obviously quite impressed by both these stories and strongly feels that they are not being given their proper due, or even worse that those who experience such numinous happenings are somehow punished for their retelling. As Kripal complains,

"As with the heads of Hercules' Lernaean Hydra, however, with every story we so decapitate, three more, or three thousand more, appear. We are swimming in a sea of such stories, if only we could recognize our situation. We do not know how many such stories there might be, much less what they might mean. We do not know because we have never really tried to find out.

Why, after all, would we study something that does not exist? 'Water?' the fish asks. 'What's water?'

It is worse than that, though. It is not just that we are told that such things, which happen all the time, cannot happen at all. It is that there are subtle, and not so subtle, punishments in place for those who take such events seriously—that is, for those who let the Hydra stand. Note that both stories feature a kind of professional fear. Twain struggled for years with whether to own his experiences in print. Even the hospital chaplain was shaken to the core by what he encountered. Clearly these events violate something basic about our worldview and our established ways of knowing. That is why Amatuzio titled her book Beyond Knowing."

Kripal's protestations seem a tad overstated here, since there is a plethora of books and articles touching upon paranormal themes that have been published both within and outside academia. The real problem, I would suggest, is that these stories are merely anecdotes told after the fact, and don't, as such, proffer overwhelming evidence to convince us that something truly extraordinary has transpired. Granted that Pierre-Simon Laplace's dictum "The weight of evidence for an extraordinary claim must be proportioned to its strangeness" can be an exacting standard, but why should we settle for less in a field famous for fostering fraudulent claims? It is not so much that extraordinary claims demand extraordinary proofs (to echo Carl Sagan and other skeptics), but that one has to be acutely cautious in differentiating that which may be genuine from that which is merely the result of smoke and mirrors. Arguably, the paranormal arena has a greater share of cranks and charlatans than any other with the possible exception of medicine and religion.

Because of this, it is vitally important to be doubly skeptical of any would-be anecdote alleging to be proof of something transcendent. For example, Kripal fails to fill in some very necessary blanks with regard to the wife's dream vision of her husband's car accident. First, what was he doing driving that late at night? Second, what are the conditions of the road near her house at that hour? Third, has her husband had any accidents of a similar nature before? Fourth, has she had any other dreams that late at night where she worried about her husband's whereabouts? These questions are just the tip of a whole slew of pertinent queries that need to be comprehensively answered before we prematurely jump from the mundane into the super mundane region of possible explanations.

I can draw from two very unusual experiences of my own which on the surface look to be of a completely inexplicable nature, particularly if one doesn't delve too deep into the contextual details, but which may not be so unusual given more, not less, information.

Back in 1981 I was teaching Psychology at Chaminade College Preparatory, a Catholic High School in Woodland Hills. One of my students came to me after class on Monday with an anguished expression on her face. She had a most terrifying dream the night before where two of her classmates (both boys and both surfers) got into a horrible car accident where they died being engulfed in flames of fire on Las Virgenes road late at night. I told her not to worry about it and just let it go as a nightmare. However, the following day (Tuesday) she had the same dream and got doubly worried. I told her again not to take it too seriously. But each night until that Friday she had the same recurring nightmare. That weekend, to the girl's shock, the two boys did indeed get into a bad car accident on Las Virgenes road confirming what she had dreamt would happen.

Given just this outline, the girl's psychic precognition would appear to be so bizarre as to resist a rational explanation. However, if the reader learns more details about the two surfers, their driving habits, and what it is like to drive with them on the weekend, it may appear less compelling as a supernatural narrative.

These two boys, who I also taught in class, were avid surfers and the one route that they took to the beach (indeed, the same route I would take on my own surfing adventures in this area) was via Las Virgenes road, since it took one directly to Malibu. This road has had an infamous history of accidents, since it has several long winding sections that veer dangerously close to jagged walled cliffs. In addition, these two surfers were well known at Chaminade to drive too fast for their own good. They would sometimes brag about their speeding efforts so as to make certain they would get waves before sunset. Moreover, even though these two boys were in a bad car accident none of them were seriously injured, which was contrary to the young girl's dream.

Yes, it was an uncanny dream, but knowing what we do about these surfers is it really a tale that requires a supernatural explanation? I think not.

Maharaj Charan Singh Ji (1916–1990)

My second story concerning a psychic nature is a very personal one for me since it involved my late guru, Charan Singh, and my teaching assistant at the time, Michelle Lopez. I was initiated by Charan Singh of Radhasoami Satsang Beas in November of 1978 and had been an avid devotee of him since I discovered him at the age of 17 (which is another curious story, best told at another time). Well, it was spring of 1990 and I was in my first year teaching Philosophy in a tenure track position at Mt. San Antonio College. One lazy afternoon, Michelle Lopez inquired about what spiritual path I was following and was keenly interested in knowing more. I hesitated, since I tended to keep my own meditational practice private and had no desire to tell others, particularly students, of my Indian guru. But Michelle was quite an unusual person who had a deep yearning for something that Western philosophy and her own Catholic upbringing couldn't fulfill. I thought it might be best to give her the brush off by having her read a book, since being a full-time student she wouldn't have time until the end of the semester to read it. The book was by Daryai Lal Kapur and entitled Call of the Great Master and dealt with the life and teachings of Sawan Singh, who was both the grandfather and guru of Charan Singh. To my amazement, Michelle came by my office the very next day saying she had stayed up all night reading the book and was anxious to write to my guru directly. I again hesitated, but looking at her existential desperation I reluctantly gave her his address at the Dera in the Punjab. She wrote a long letter to him that night.

Several weeks late I was awakened very early in the morning by a telephone call from my good friend Paul Tooher who told me the heart wrenching news that Charan Singh had died. This was June 1, 1990. I was devastated since I was quite attached to him. Later that morning I had to take a train from Del Mar to Anaheim (where I kept a car) so as to make it to classes that I wanted to cancel but which I nevertheless attended. However, before I entered into the classroom, I walked over to my office and was surprised to be met by Michelle Lopez who seemed extremely distraught. I asked Michelle why she was there so early and why she was so upset. She then told me the most remarkable story of how she was driving the night before on the 405 freeway and had to pull off on the side of the road because she was overwhelmed by a vision of light. In that light she learned that my guru, Charan Singh, had died. She couldn't believe it and so came to see me early because she was hoping it was just a massive hallucination. When I told Michelle that Charan had indeed died, she was overcome with emotion and couldn't stop sobbing. I too was taken aback, since the time she had her vision and the time that Charan died were spookily correlated.

I asked Michelle if she ever got a reply to her earlier letter to Charan Singh and she said nothing had shown up. I explained that he probably didn't write because he was too busy to respond since he has over a million initiates in India alone.

However, two weeks later at the Mt. San Antonio commencement ceremonies, Michelle came running up to me and said that she had just gotten a letter from Charan Singh that morning. Two weeks after his death? That seemed impossible. But it turned out that Charan had indeed dictated his reply to Michelle but had died before he could sign the letter that was forwarded to her with a note explaining the unusual circumstances. Personally, it was a very sweet letter for me since he had mentioned my name in the body of the letter and told Michelle to seek my advice on understanding the philosophy of Sant Mat.

How can such a story be relegated to the dustbin of mere chance and coincidence? Isn't Kripal correct when he cautions us to transcend our scientism and open our minds up to alternative, even supernatural, explanations? As Kripal implores, "I suggest a way out of our present impasse: We should put these extreme narratives, these impossible stories, in the middle of our academic table. I would also like to make a wager, here and now, that once we put these currently rejected forms of knowledge on our academic table, things that were once impossible to imagine will soon become possible not only to imagine but also to think, theorize, and even test. I am betting, in other words, that we actually need these so-called impossible things to come up with better answers to our most pressing questions, including the biggest question of all: the nature of consciousness."

Kripal is, of course, correct in his advice that we put these narratives straight and center on our academic agendas, but I think it is already happening much more than he may wish to acknowledge. It just so happens that when we put such narratives under skeptical microscopes much of their magical spell dissipates and gets replaced with the weight of more ordinary factoids.

The real difficulty, I would argue, is being critically analytical of our own numinous or borderline encounters.

The real difficulty, I would argue, is being critically analytical of our own numinous or borderline encounters. Richard Feynman, the distinguished physicist at Cal Tech, rightly cautioned years ago "The first principle is that you must not fool yourself and you are the easiest person to fool."

Looking more objectively and with an eye for untold details concerning Charan Singh's death and Michelle's extraordinary vision, could there possibly be a more sedate explanation for what occurred?

I think so, even if believers will still hold on to its seeming paranormal implications. First, was there any reason for Michelle to believe (prior to her vision) that Charan's life expectancy was coming to a close? Yes. I, myself, had long worried that he would not live much longer. I thought this not because I was psychic but because I saw some telltale signs in Charan's own behavior, not the least of which was his rushed publication, *Treasure Beyond Measure*, that revealed for the first time in print very personal details about how he assumed the mastership and his frank admissions about feeling unworthy of such a position. He also opened up the Dera colony to all visitors from around the world, something he had never done before. I imagine that I would have conveyed some of my thoughts about Charan's unprecedented actions to Michelle. Furthermore, Michelle had sent her letter to Charan sometime in April and given his legendary punctuality she may have wondered about the absence of a more timely response. Interestingly, the night that Michelle had her vision of Charan's demise, she had just attended a spiritual talk by a woman guru from India and was driving home when she was overwhelmed by her spiritual revelation.

I realize that elaborating on these contextual details may be unsatisfactory, but it does underline an important issue inherent in these supernatural recollections: the more information we know about them the better informed we will be about adjudicating their alleged transpersonal veridicality.[3]

36

To Jeffrey Kripal's credit, he readily admits that paranormal stories are not untainted or virginal in their retelling. As Kripal telling reveals, "It is not just our fault, though. There are fundamental ambiguities inherent in the experiences themselves, ambiguities that make it difficult to put and keep these experiences on our academic tables. To start with, these things are not things. Nor are they replicable or measurable. And then there is the key role that the human imagination plays in these visions. . . . Finally, the recounting of even the empirical cases is often changed in small ways (missing an important detail or supplying a nonexistent one), which suggests that these visions are accurate anomalous cognitions that have been 'filled in' with imagined details—mixtures of trick and truth."

I think Kripal is more on point when he sides with the correctness of early-Victorian researchers who "called dreams like the two with which I began 'veridical hallucinations,' or hallucinations corresponding to real events."

But what Kripal seems resistant to accepting is how amazing synchronicities and the like are actually destined to happen given the laws of probability and the theory of large numbers. Indeed, even small random numbers can generate unexpected results, as has been amply demonstrated time and again by the famous birthday paradox where the odds that two people having the same birthday is much higher than one would expect, even with a gathering of just 35 people.

Get a large enough number set and just about any crazy coincidence is bound to happen, as J. E. Littlewood from Cambridge University suggested in the previous century. His work in this regard has been popularly entitled by Freeman Dyson at the Advanced Institute at Princeton as "Littlewood's Law of Miracles" which basically states that after one million or so random events (and if we are open enough to the probability matrix that surrounds them) we should expect a miracle! Littlewood calculated that we would experience about a million events every 35 days, so that the odds are in our favor to witness nearly 12 miracles a year.

In earlier essays I have developed a variation of Littlewood's Law called *Desultory Decussation* (where two apparently random events intersect to form an X): "If there are thousands, nay millions, of events in our lives (measured in transparently fractal ways), then it should be expected that for every 10,000 plus events, there may be two or more events which intersect. Notice that intersection and you will be aware of a meaningful coincidence--the meaning being that two disparate parts have something in common (whatever that intersection may entail).

We can even splinter off from this and make a broad sweeping generalization. There are those who look or seek out these desultory decussations and those who do not. I would imagine that some of us are more attuned or keenly aware of the intersections (which happen randomly) and they will end up seeing more meaning in their lives, even if the meaning quota is the same relatively speaking for all.

In other words, there are those who seek the Littlewood stream and plunge right in and those who do not. Blind typing may in fact produce a legible word just by chance, but the key in all this is to actually become aware of that probability and notice it when such does occur. Otherwise, so many amazing happenings of chance go by completely undetected. If we could remain conscious of this mathematical matrix, we could be experiencing stunning hierophanies not only monthly, but perhaps daily. We already know that the theory of large numbers bears this possibility out. The only real glitch resides within our selves. To experience Littlewood miracles or desultory decussations (random events interwining in meaningful X patterns), it takes a Herculean effort on our part to remain open to what strange coincidences nature may throw out at us. Littlewood's Law, interestingly enough, first requires us to be attentive, exceptionally so."

David J. Hand, Senior Research Investigator and Emeritus Professor of Mathematics at Imperial College, London, and Chief Scientific Advisor to Winton Capital Management has written a book whose title captures how truly unusual moments occur all the time, *The Improbability Principle: Why Coincidences, Miracles, and Rare Events Happen Every Day*. As Hand explains his thesis,

"The improbability principle is composed of five laws, analogous to the four laws of thermodynamics or Newton's three laws of motion. These laws, the law of inevitability, the law of truly large numbers, the law of selection, the law of the probability lever, and the law of near enough, explain exactly why we should expect to encounter highly unlikely events, and indeed why we should expect to do so on a regular, even frequent, basis. Any one of the laws acting by itself can lead to a highly improbable event—like people winning the lottery twice, or 26 black numbers coming up one after another in roulette."

Yet, Kripal's long essay in the Chronicle is not merely about the study of improbable occurrences, but more pointedly a critique of a purely materialist agenda that he feels has saturated and (to some measure) poisoned the well of academia to such an extent that religious scholars cannot even take the core of their subject—the religious—seriously. As Kripal tellingly argues in a long passage:

"In the rules of this materialist game, the scholar of religion can never take seriously what makes an experience or expression religious, since that would involve some truly fantastic vision of human nature and destiny, some transhuman divinization, some mental telegraphy, dreamlike soul, clairvoyant seer, or cosmic consciousness. All of that is taken off the table, in principle, as inappropriate to the academic project. And then we are told that there is nothing 'religious' about religion, which, of course, is true, since we have just discounted all of that other stuff. Our present flatland models have rendered human nature something like the protagonist Scott Carey in the film The Incredible Shrinking Man (1957). With every passing decade, human nature gets tinier and tinier and less and less significant. In a few more years, maybe we'll just blip out of existence (like poor Scott at the end of the film), reduced to nothing more than cognitive modules, replicating DNA, quantum-sensitive microtubules in the synapses of the brain, or whatever. We are constantly reminded of the "death of the subject" and told repeatedly that we are basically walking corpses with computers on top—in effect, technological zombies, moist robots, meat puppets. We are in the ridiculous situation of having conscious intellectuals tell us that consciousness does not really exist as such, that there is nothing to it except cognitive grids, software loops, and warm brain matter. If this were not so patently absurd and depressing, it would be funny."

But does Kripal's materialist summation really capture what is happening across scientific disciplines? I think not. Yes, to be sure, there are those who fit into Kripal's caricature, but one can readily think of many more scientists and researchers that do not. M.I.T., for instance, one of the premier "materialist" institutions in the world has published a whole line of monographs and books exploring how eastern forms of meditation (Zen or otherwise) connect with current studies in neuroscience. The University of Arizona has sponsored all sorts of studies on consciousness for the past two decades, allowing for a wide variety of contributors—many of whom favor distinctly non-materialist understandings of self-reflective awareness. Even Harvard University, the gold standard of all things scientific, has featured a slew of books focusing on the central importance of consciousness and which do not dismiss its central importance.

Perhaps the real thorn in Kripal's claw is the diminishing impact that the humanities in general have in shaping intellectual discourse today, especially as the hard sciences have emerged as having a much greater influence and say among today's technological savvy audience. As Kripal confesses,

"Humanists have paid a heavy price for their shrinking act. We are more or less ignored now by both the general public and our colleagues in the natural sciences, whose disciplines, of course, make no sense at all outside of universal observations, and who often work from bold cosmic visions, wildly counterintuitive models (think ghostlike multiverses and teleporting particles), and evolutionary spans of time that make our 'histories' look insignificant and boring by comparison."

I think Kripal's worry is misguided and based on a fundamental misreading of what materialism portends.

But I think Kripal's worry is misguided and based on a fundamental misreading of what materialism portends. Indeed, as I argued in a filmed keynote presentation at the International SPIRCON conference held at the Dayalbagh Educational Institute in Agra, India, four years ago, the real impasse between religion (or, in this parameter, the humanities) and science stems largely from a linguistic conundrum over what the word "matter" ultimately means. Kripal seems under the mistaken impression that relegating everything to a material basis somehow negates or eliminates the mystery of the universe at large and renders useless any cosmic feeling of importance that we may once have had. To use Kripal's metaphor, we have been shrunken almost to the point of oblivion by our overly reductionist purviews. Kripal uses Scott from the Incredible Shrinking Man as a clarion call against such materialist tendencies, but in so doing seems to have forgotten the most moving part of the entire movie wherein Scott realizes that no matter how much he may shrink and no matter how small he may become in comparison to others, the vast mystery of the universe doesn't lessen one iota but becomes ever greater as he appreciates that the infinitely small and the infinitely large meet. It is through his incredible shrinking that he gains a much deeper spiritual insight into the vastness of the cosmos of which he is intimately linked.

There is a wonderful majesty to the multiverse we find ourselves in and it isn't lessened by science's naturalistic methodologies in the least. To the contrary our materialist science has opened vistas unimagined by our ancestors by focusing precisely on that which can indeed be observed by our increasingly sophisticated sensory extensions—be they microscopic or macroscopic in their reach. What is more mind blowing? The Genesis account of creation (where God creates two lights in the sky, one to the guide the day and one to guide the night, apparently not realizing that the moon isn't a source of light as such, but rather a reflector of the sun) or the latest findings in astrophysics, where almost weekly new discoveries are being made which tell us about black holes a million times larger than our sun, vast galaxies with innumerable stars and planets, and strange mysteries such as dark matter and dark energy? Astronomy is Genesis rewritten and expanded nightly.

No, the humanities aren't shrinking in the least but it is rather taking on a new and richer form, the likes of which can only increase our appreciation of human culture and ingenuity. The new humanities are one that is not distanced from the hard sciences (as C.P. Snow famously explained in his lecture about the two cultures) but is intimately connected with them, and because of this is offering up stunningly beautiful and novel pathways to better understand humankind and its place in the cosmos.

Simply put, I don't see Kripal's alleged impasse between matter and spirit. I see, rather, a persistent confusion on our part for not taking matter and what it implies seriously enough. For if we do that, and don't hide our heads in the sand with unnecessary and outdated dualisms, we soon realize that matter is just as mysterious, if not more so, than anything described in our spiritual literature. Matter is multi-dimensional and doesn't in itself exclude anything whatsoever in the human experience, but rather embraces everything just as an ocean includes even the tiniest of water drops and the largest of tsunamis.

This doesn't mean to indicate that we ultimately know what matter is. We don't and therein lay the great truth that upends those with a religious bent who are afraid of being reduced to mere physicality. Saying we are merely this body, merely this stuff doesn't lessen who we are one bit. Why? Because there is nothing "merely" about matter, since the word itself is but a placeholder for "something doing something that we ultimately do not know what," to bastardize a famous line from the astronomer Sir Arthur Eddington.

Our consciousness won't be lessened if it turns out to be an emergent property of the brain and its 86 billion neurons, just as the ocean isn't lessened by knowing that its fundamental parts are a mixture of Oxygen and Hydrogen. This is why I find Kripal's attempt to salvage consciousness from its materialist underpinnings to be premature and unnecessary. He mistakenly believes that we need to divorce consciousness from its material corpus (thus he invokes the often-utilized "awareness as radio signal" which moves through hardware but which is not created by it) so as to more properly explain such anomalies as Twain's dream and Michelle's inner vision. But this, as I have already pointed out isn't necessary if indeed we live in a vast probability matrix, the likes of which can on occasion (just by the vastness of numerical intersections and life experiences) bring about the most startling and unexpected of synchronicities. Imagine how many precognitive dreams we have had in our lifetime and how many of them have never amounted to anything significant. Add some 7 billion dreamers to that mix and start calculating the odds that some dreams may stand out as remarkable predictors of future events. Following Littlewood's lead and David Hand's Improbability Principle and we should expect a lot more Twain-like precognitions than we can at first imagine. None of this necessitates believing that consciousness has to be immaterial, keeping in mind the very necessary caveat that even if awareness is purely physical it doesn't by definition mean we have truly understood it nor have we eliminated its majestic mystery.

Metaphorically speaking, I don't see why we have to invite Plato back to academia (as Kripal implores) if he has never left the building in the first place. Science welcomes all voices, even the most hyperbolic, but to be properly heard and appreciated it is necessary not to confuse innumeracy with evidence. As Plato cautioned his students so many centuries ago who wished to study at his famous academy, "Let none ignorant of geometry enter here."

Or, to summarize what Kripal may have underplayed in his anti materialist primer for embracing the unexplained, "The high improbability of an event oftentimes blinds us from the probability, even if rare, that such events are probabilistic.

NOTES

[1] I must confess that I was taken aback by Jeffrey Kripal's credulity when I read his unabashed fawning of the late cult guru, Adi Da (aka Franklin Jones, Bubba Free John, Da Free John, et. al), in his foreword to the updated edition of The Knee of Listening. It gives one pause when a scholar of Kripal's rank can be taken in by the likes of Adi Da whose track record with certain female devotees was nothing less than unconscionably abusive. Professor Scott Lowe from the University of Wisconsin and I have written a critical analysis of Adi Da's life and work nearly 20 years ago. The book is entitled *DA: The Strange Case of Franklin Jones*. Dr. Lowe was a former member of Da's group and has given a harrowing portrayal of his former guru.

[2] When I was a teenager in high school my friend Charles and I planned our first surf trip to San Clemente and the night before I had an exceptionally vivid dream of a blonde hair girl who worked at a burger joint in San Clemente, even though I had never been there before in my life. That next day when we arrived in San Clemente after surfing at San Onofre, we stopped by a drive-on hamburger stand off the freeway and when I ordered my food I immediately noticed a blonde hair girl behind the counter who looked exactly like the girl I had dreamt about the night before. I was so shocked by her appearance that I even told her about my uncanny dream. Nothing came of it, except perhaps a self-conscious smirk from her and her co-workers. Of course, dreaming of a blonde hair girl before a surf trip may not be so unusual amongst teenage boys who surf.

[3] Michelle Lopez died just a couple of years after Charan Singh's death. She was a very kind and warm-hearted person whose life was tragically cut short by a drunk driver who ran a red light. Michelle had told me long before her death that she feared she would die young and that it would most likely be from a car accident. When I asked her why, she told me of a strange psychic she had met who told her that she would die unexpectedly from a car accident. I was irritated and quite shocked that a psychic would tell her such a terrible thing, but Michelle didn't seem troubled by the prediction. Thankfully, Michelle had a beautiful baby girl before she died and right before the accident she had dropped her daughter off with relatives. All those who knew her to this very day miss her.

"Now if you suppose that there is no consciousness, but a sleep like the sleep of him who is undisturbed even by dreams, death will be an unspeakable gain. For if a person were to select the night in which his sleep was undisturbed even by dreams, and were to compare with this the other days and nights of his life, and then were to tell us how many days and nights he had passed in the course of his life better and more pleasantly than this one, I think that any man, I will not say a private man, but even the great king will not find many such days or nights, when compared with the others. Now if death be of such a nature, I say that to die is gain; for eternity is then only a single night."
— *Socrates*

In deep sleep there are no thoughts, and there is no world. In the states of waking and dream, there are thoughts, and there is a world also. Just as the spider emits the thread (of the web) out of itself and again withdraws it into itself, likewise the mind projects the world out of itself and again resolves it into itself. When the mind comes out of the Self, the world appears.
— *Ramana Maharshi*

Any organism that develops a higher form of consciousness which can reflect on its past and future suffers a most curious fate.

I liked Elliot Benjamin "Life, Death, Meaning, and Purpose" very much because it is a brutally honest essay about his existential angst when contemplating about how life may have no ultimate meaning or purpose. As Benjamin readily admits when confronted with a purely materialist universe, "this perspective leaves me feeling somewhat depressed when I think too much about it." Yes, I think most of us deep down inside have an almost built-in resistance to the idea that this life as we are living now will come to an end with no hopes of something beyond it.

It may well be that any organism that develops a higher form of consciousness (deep self reflection, extended virtual simulation, Edelman's 2nd nature), which can reflect upon its ancestral past and project far into an imagined future, suffers a most curious fate. Being able to self reflect and self project is of a great evolutionary advantage over those organisms that are stuck within the confines of instinct and an extended present moment, since it provides a plethora of options that can be simulated within one's skull before being parlayed in a real and dangerous world. Simply put, reflecting before projecting is a wondrous survival tool, except that it carries an unforeseen downside. To the degree that I am freer than other animals to directionally ponder in my mind unimagined scenarios, it allows me to better adapt to unexpected future outcomes. But this same freedom also opens me to imagine my own cessation and those of others that I am attached to and love.

A dog doesn't appear to meditate on how distant stars will die and transform into black holes, or how universes may appear and disappear over eons of time. Yet, as humans with enlarged brains we evolved to ponder all sorts of imponderables, and thus our Darwinian gift is also at times our Darwinian curse.

This is why Benjamin laments that he may lack a gene (or an added one), since he cannot quite comport with a purely materialist worldview, "Sometimes I wonder if I lack a gene, or perhaps have an extra gene, that makes me 'different' from other people."

Benjamin is not alone in his sentiment nor do I think his genetic predispositions are different than most. I too feel as Benjamin does, but that is precisely why I think a deeper understanding of how consciousness operates can, to some measure, liberate us from our 2nd nature nausea (to slightly bastardize a famous observation from Albert Camus).

The seemingly most important questions we tend to ask (Is there a God? Do I have a soul? Is there life after death? Etc.) only arise at certain moments in our awareness and completely subside at other moments. Indeed, within any normal 24 hour cycle, the issues we think that are so vital and so urgent only last for a set duration only to disappear when deep sleep overwhelms us and all such questions are forgotten until we awake again.

In a sense, the questions we ask of the universe are really reflections of our state of awareness at any given stage and therefore serve as telling signposts of what that particular state entails. The questions and their desired answers are of secondary importance when viewed in this cyclical light. What seems more elemental is to find out precisely why these pressing concerns bother us so much at 12 noon, but appear lifeless and without urgency at 12 midnight when we are fast asleep.

The thesis is a very simple, if a profound one: We do not fear death in deep sleep and we do not care if there is life after death or if there is a God or who we are. These issues only arise within a certain frame of consciousness which itself is but a temporary state. Change the state and you change the questions that appear to matter so much to us. Our day-to-day experiences are truly one of consciousness interruptus: we grog awake, we stay awake, we take naps, we space out, we grog to sleep, we dream, we fall into a deep slumber. And the process recycles. Depending on which cycle is operative, so too the questions we ask and ponder. We live in a temporal context, a self-referential trap, and hence our first error is that we confuse neurology or our present state of awareness with the absoluteness of all things, forgetting that our awareness is akin to *Frijoles saltarines*, Mexican jumping beans: going from place to place, but never stationary, never permanent. Just as waves modulate upon the sea, so do our moments of self rise and fall.

I am a strange loop, as Douglas Hofstadter rightly surmised in his book of the same title, but which he first introduced in *Gödel, Escher, Bach: an Eternal Golden Braid*. Depending on which state of awareness I find myself, the feedback circuits alternate and so do my concerns. Most of the time when I am in the waking state I am not doing philosophy, but instead focusing on a whole slew of bodily needs, which demand my attention . . . from eating to drinking to showering, etc.

And no one state remains constantly the same but fluctuates throughout the day and night, giving rise to a cascade of moods and feelings. Yet, each of these states of mind have their own temporary context and instead of one permanent self throughout these changing proceedings I find instead a multiplicity of mini selves each with their own unique perspectives, desires, and troubles. Yes, there is a part of me that attempts to stich all these personas into a unified person I call "my" self, but this unconscious (and sometimes conscious) activity continually gets interrupted. At one moment I may be fearful about flying and what it portends, particularly if I learn about an airline tragedy as happened these past few weeks with the disappearance of a Malaysian jet. At another instance, I don't care about anything else except the tasty lemonade I am sipping and the glistening sun shimmering off the trees in Idyllwild.

Perhaps our consciousness can be likened to ocean waves which, depending on their size and shape, we ride in different ways. A wave at Point Panic on the south shore of Oahu necessitates that I bodysurf it going right, with only fins on my feet as my hands plane through its bowling section. However, if I am surfing at the north side of Newport Pier in the dead of

Winter I may opt for a 9-foot long board since on occasion I am forced to navigate the barnacled pilings to safely make my way to the other side. Or, if I am staying at La Jolla shores and playing with my two kids and my wife, we may all use body boards and ride white water together until we land on dry sand.

Each wave is different and therefore allows for different possibilities. Analogously the same holds true for the undulating ripples of our own awareness. This is important because quite simply the context of consciousness shapes and contours the content of what arises and what necessitates our attention. The questions we ask of ourselves and of the multiverse at large are products of these changing patterns, whether they are environmentally cued or biochemically triggered.

Thus, I have long noticed that whenever we are happy—very happy—at a particular juncture in life we may also feel a certain anxiety, a certain fear that it may end. Yet, if we are deeply depressed or extremely ill, we don't worry as much (if at all) about death or non-existence. Therefore our questions and worries are a priori predicated upon our wavering moods. Instead of getting trapped with our self-referential feedback loops, pondering conundrums that may never be resolved, it might be wise to introspect on why certain questions only emerge at certain times and not at others. In other words, instead of a continual stream of unanswered queries, we focus on why we are asking why in the first place, such as why I fear death only when I am relatively happy but never when I am severely sick with stomach flu. The questions we ask have less to do with some ultimate truth but more to do with our own neurological phases. This is telling because if we could better understand the ground from which our varying forms of awareness arise, it may transform our reactions when these strange loops of awareness take shape.

Apparently sages from both East and West realized the volatile nature of consciousness and instead of being sabotaged by its manifold incarnations, sought to seek its phenomenal origins. Hence, the non-dual philosophy of Indian mystics such as Shankara or Ramana Maharshi becomes clearer sense when self-awareness is viewed in this light. The following question and answer sequence, transcribed live from a conversation between Ramana and a seeker at Arunachala ["Talks with Sri Ramana Maharshi"], provides a glimpse into how such an enlightened insight can transfigure how we view ourselves and our surrounding cosmos.

D.: Should we not find out the ultimate reality of the world, individual and God?

M.: These are all conceptions of the 'I'. They arise only after the advent of the 'I'-thought. Did you think of them in your deep sleep? You existed in deep sleep and the same you are now speaking. If they be real should they not be in your sleep also? They are only dependent upon the 'I'-thought. Again does the world tell you 'I am the world'? Does the body say 'I am body'? You say, "This is the world", "this is body" and so on. So these are only your conceptions. Find out who you are and there will be an end of all your doubts.

D.: Being always Being-Consciousness-Bliss, why does God place us in difficulties? Why did He create us?

M.: Does God come and tell you that He has placed you in difficulties? It is you who say so. It is again the wrong 'I'. If that disappears there will be no one to say that God created this or that. That which is does not even say 'I am'. For, does any doubt rise 'I am not'? Only in such a case should one be reminding oneself 'I am a man'. One does not. On the other hand, if a doubt arises whether he is a cow or a buffalo he has to remind himself that he is not a cow, etc., but 'I am a man'. This would never happen. Similarly with one's own existence and realisation.

Another from the group asked: How is the ego to be destroyed?

M.: Hold the ego first and then ask how it is to be destroyed. Who asks this question? It is the ego. Can the ego ever agree to kill itself? This question is a sure way to cherish the ego and not to kill it. If you seek the ego you will find it does not exist. That is the way to destroy it. In this connection I am often reminded of a funny incident which took place when I was living in the West Chitrai Street in Madura. A neighbour in an adjoining house anticipated the visit of a thief to his house. He took precautions to catch him. He posted policemen in mufti to guard the two ends of the lane, the entrance and the back-door to his own house. The thief came as expected and the men rushed to catch him. He took in the situation at a glance and shouted "Hold him, hold him. There-he runs-there-there." Saying so he made good his escape. So it is with the ego. Look for it and it will not be found. That is the way to get rid of it.

M.: Who is this witness? You speak of 'witness'. There must be an object and a subject to witness. These are creations of the mind. The idea of witness is in the mind. If there was the witness of oblivion did he say, 'I witness oblivion'? You, with your mind, said just now that there must be a witness. Who was the witness? You must reply 'I'. Who is that 'I' again? You are identifying yourself with the ego and say 'I'. Is this ego 'I', the witness? It is the mind that speaks. It cannot be witness of itself. With self-imposed limitations you think that there is a witness of mind and of oblivion. You also say, "I am the witness". That one who witnesses the oblivion must say, "I witness oblivion". The present mind cannot arrogate to itself that position.The whole position becomes thus untenable. Consciousness is unlimited. On becoming limited it simply arrogates to itself the position. There is really nothing to witness. IT is simple BEING.

Sam Harris in a recent interview ["Taming the Mind"] with Dan Harris of *ABC's Nightline* (apparently not related) gave a pregnant explanation of what meditation accomplishes (or doesn't accomplish?) by focusing on the illusion of a self: "The same is true for the illusoriness of the self. Consciousness is already free of the feeling that we call "I." However, a person must change his plane of focus to realize this. Some practices can facilitate this shift in awareness, but there is no truly gradual path that leads there. Many longtime meditators seem completely unaware that these two planes of focus exist, and they spend their lives looking out the window, as it were. I used to be one of them. I'd stay on retreat for a few weeks or months at a time, being mindful of the breath and other sense objects, thinking that if I just got closer to the raw data of experience, a breakthrough would occur. Occasionally, a breakthrough did occur: In a moment of seeing, for instance, there would be pure seeing, and consciousness would appear momentarily free of any feeling to which the notion of a "self" could be attached. But then the experience would fade, and I couldn't get back there at will. There was nothing to do but return to meditating dualistically on contents of consciousness, with self-transcendence as a distant goal.

However, from the non-dual side, ordinary consciousness—
the very awareness that you and I are experiencing in this
conversation—is already free of self. And this can be pointed out
directly, and recognized again and again, as one's only form of
practice. So gradual approaches are, almost by definition,
misleading. And yet this is where everyone starts.

In criticizing this kind of practice, someone like Eckhart Tolle
is echoing the non-dualistic teachings one finds in traditions
such as Advaita Vedanta, Zen (sometimes), and Dzogchen.
Many of these teachings can sound paradoxical: You can't get
there from here. The self that you think you are isn't going to
meditate itself into a new condition. This is true, but as Sharon
says, it's not always useful. The path is too steep.

Of course, this non-dual teaching, too, can be misleading—
because even after one recognizes the intrinsic selflessness of
consciousness, one still has to practice that recognition. So there
is a point to meditation after all—but it isn't a goal-oriented one.
In each moment of real meditation, the self is already
transcended."

The virtual simulator theory of consciousness basically
suggests that there is no absolute concrete self, as such, but
rather a kaleidoscope of enveloping simulations which provide
us with an almost infinite array of stratagems for future
orientations or past musings for what was or could have been.
Rarely do we ever experience the space prior or between such
mental facsimiles. Yet this very ground of being is always
present and never absent from any phrenic permutations, just as
in movies which may change nightly whereas the projector and
the screen upon which they are projected remain constant.

Realizing that consciousness unfolds in this sequential
manner (with dreaming being the most illustrative example of
how a virtual simulator operates) it provides us with a bridge
with which to better appreciate Benjamin's sorrow when he
writes, "But if all I am is a formation of chemicals that came
about for no 'meaningful' or 'purposeful' reason whatsoever, and
when I die that is it—well to be very candid I find this sad."

This is what meditation can unlock: direct insight into how
the mind constantly seduces us into believing its conjurings to be
the sum total of reality.

Yes, it is indeed sad, but only during a very specific stage in
awareness. Such sadness doesn't arise in us when we are fast
asleep or when we are slurping on a classic Coke and not
thinking about much of anything except that next bite of a hot

pretzel dipped in mustard. This is not to dismiss Benjamin's astute observation about meaninglessness, because we all share (more or less) that same sense of dread at certain points, but to dig deeper and find out in what specific contexts does our individuated selves loom so large as to bring such angst into the forefront of our attention. This is what meditation can unlock: direct insight into how the mind constantly seduces us into believing its conjurings to be the sum total of reality, when in, point of observation, they are merely mapping transparencies evolved over long gestations of time to help 2nd nature organisms better survive and adapt in this Hunger Games like carnivore show we call living. Look to the source of where such forms of consciousness emerge and follow how long these temporary aberrations persist and the Ferris wheel nature of self-reflective awareness becomes all too obvious. We may not be able to stop our mind from its simulating habits (after all, developing such a turnstile of past and future ruminations is of a great evolutionary advantage to us), but if we can remain a witness and not a prisoner to its concealing allurements, then we can be liberated from its more negative consequences. In this way, we can better enjoy the firework display of consciousness because we are keenly cognizant of its ephemeral nature and purpose.

Of course, this doesn't mean that by our Ramana like inquiries (or meditations) we have somehow solved the riddle of existence and forever ended the samsara of pain, which inevitably arises within the human condition, but it does provide a profound pathway to alleviate much of the unnecessary sufferings we experience because we have conflated our current brain state for the "real" state of the universe.

Benjamin brilliantly captured the essence of the virtual simulator and its machine like ability to churn over unsolved mysteries when he wrote, "Well it's Friday night and I think I'll indulge in my frequent Friday night habit of spacing out with some wine cooler and cheese doodles while I listen to some very old-fashioned show music from the 1950s on my 'record player.' But while I am doing this, my subconscious will be hard at work pondering if there could possibly be any meaning or purpose to the universe, and if I come up with any creative insights I'll be sure to let you know."

The good news here is that as Ramana Maharshi and other Advaita Vedantists may be quick to point out, much later in the night when we are all in a dreamless sleep, we will remain blissfully unaware not only of this world and ourselves, but also of worries about meaning and death and the purpose of it all. As Plato opined (via the mouthpiece of Socrates) centuries ago in his Apology, "Now if you suppose that there is no consciousness, but a sleep like the sleep of him who is undisturbed even by dreams, death will be an unspeakable gain."

The secret is to understand that the questions that mean so much to us when fully awake (and which serve as relative thermometers of our own well being) lose all their compelling force when we enter a different region of awareness. Awaken to this evidential truth, and the existential questions that have driven humankind near the brink of madness since time immemorial melt away like ice in an unremitting desert sun.

"To mistake a wave (and what it brings forth) with the totality of the ocean is like confusing a state of awareness (and its implications) with the ultimate reality of all that exists."
—*The Lost Manuscript of Theodore Wilkins*

"Do not be attached to the phantasms of light and sound and the magical universes they create, but ask a deeper question. Who is it that sees this light within? Who is it that hears the inner sound? Find the source from where such light and sound flows forth. Find the source of why so many questions arise from this I' thought. That source is our real being."
—*Baba Faqir Chand (free translation from a personal letter dated 1980)*

I appreciated Elliot Benjamin's agnostic purview as outlined in his recent article ["Agnosticism and Fundamentalist Mediumship"], particularly his acceptance of alternative viewpoints concerning how mediums garner information. I found the unnamed "President's" pointed rejoinder to Benjamin (where he snidely queries, "do you think that everything Edgar Cayce said in his readings was just his imagination?") to be not only silly, but also more of a rhetorical boomerang than he might have realized. Why? Because a close analysis of the Cayce readings that was conducted by K. Paul Johnson, a sympathetic scholar who was encouraged in his research by A.R.E., didn't find anything of a truly paranormal nature.

I had the opportunity of reading Johnson's manuscript before it was published by SUNY Press and found it to be a remarkably objective study—so impressive, in fact, that I wrote a long review of it, which I include here since I believe K. Paul Johnson and Elliot Benjamin share much in common. Perhaps if that unnamed "President" closely studied Edgar Cayce in Context he would realize why Elliot Benjamin is spot-on when he writes, "But I'll say again that my agnostic perspective looks at all possibilities, and the interpretation of coming from my imagination for what I experienced at my Mediumship workshop is a very reasonable one, and I know the process I was using at the time and it felt to me very artificial, and there are all kinds of interpretations that do not involve either an afterlife or psychic interpretation for anything I heard in the responses from people that I worked with at my conference."

A REVIEW OF EDGAR CAYCE IN CONTEXT

In this new study of Edgar Cayce, Mr. Johnson exemplifies a most remarkable methodological bias: open-mindedness. Unlike typical studies of the paranormal or the transpersonal, where the reader is left with either adopting a believer's or a skeptic's position, Johnson weaves an illuminating pathway by which one can see where Edgar Cayce's readings have been historically and factually inaccurate or where they indicate a potential transrational imperative. In either case, Johnson is masterful in avoiding the pitfalls that usually sink investigations of this kind.

As he demonstrated with his pioneering books on the history of the Theosophical Masters, where he literally grounded the metaphysical Great White Brotherhood down into the social and political moorings of the late 19th century, Johnson places Edgar Cayce in the larger, infusing environment of the early and mid-20th century. By doing so, the reader begins to appreciate the religious context out of which Edgar Cayce was operating, and, in turn, how to better appraise the import of his trance readings.

Johnson does not believe he has the final answer on Cayce's psychic abilities. Rather, he has taken a multi-dimensional view of the man and in so doing can easily navigate between the waters of empiricism and occultism, while all along remaining relatively unscathed and still objective. Undoubtedly, Johnson is the best guide we have on Edgar Cayce to date.

Having said that, however, my task is not to adopt Johnson's broader phenomenological perspective, but rather to illustrate how a skeptic, particularly one steeped in western science, grapples with the phenomenon of Edgar Cayce. In other words, my task is to "explain" the apparently transpersonal or paranormal elements suggested in the Cayce readings. My approach is decidedly reductionistic (a term I use unhesitatingly and with approval) and therefore tends to look for the simpler, more earthy interpretation of any paranormal claim, whether it be in the realm of ufology, medicine, astrology, or psychic gifts

Thus, in the case of Edgar Cayce's trance readings I have employed the principal tool of my trade: Occam's Razor. Essentially what this entails is "shaving" down the extraordinary claims surrounding Cayce's readings and attempting to discover a more ordinary explanation. Now Occam's Razor is not a magic blade and it should be remembered that it doesn't always work. It so happens that some phenomena are not quantifiable or reducible. They resist wholesale reductionism and must be understood on entirely new levels of explanation. It may be true to say that a futuristic novel, like Aldous Huxley's Island, is ultimately comprised of letters, 26 individually distinct symbolic units, but entirely misleading to suggest that reading those components in isolation is all that is necessary to understand Huxley's meaning and intention. Obviously, the novel must be read in its entirety (from whole sentences to whole chapters to the whole book) in order to properly appraise all of its various facets.

Thus the spirit of reductionism is not to deflate everything no matter what to its atomic structure, but rather to simplify and reduce those things that are amiable to such reduction. To say that water is H20 is illuminating, since we get a deeper insight into how and why water is formed. It is what philosophers of mind, like the Churchlands, call intertheoretic reduction, an entirely appropriate and meaningful way to grapple with physical mysteries. But to say that the Encyclopedia Britannica is nothing more than alphabet manipulation is to completely miss its most important feature: information. Such information, though comprised of smaller units (whether they are comprised of English, French, or binary), cannot be comprehended until its higher levels of organization are ascertained and understood: the word, the sentence, the paragraph, the page, the chapter, etc. It is on those higher levels where the fullness of information conveyed in the encyclopedia can be appreciated.

But keep in mind one important caveat: regardless of how sophisticated, or higher level order, our information may be—as in the case of the printed or online versions of the Encyclopedia Britannica—it is always built algorithmically: one step, two step, three steps; A, B, C; one letter, one word, one sentence, one paragraph, etc. Our world is a scaffolding project and the closer we pay attention to the various steps inherent to that scaffolding the more accurate and precise our descriptions of the universe become.

How this relates to the field of parapsychology in general and to Edgar Cayce specifically is twofold: 1) before we entertain theories that are trans-rational we should attempt to discover explanations which are rational or pre-rational; and 2) if it so happens that no adequate scientific explanation is possible, even with our current state of advanced technology, we should not succumb to ad hoc transcendental theorizing. Why? Because the very moment we opt for "sky hooks" (Daniel Dennett's lovely phrase for non-algorithmic guesses in contradistinction to "cranes" which are algorithmic and procedural) we have, more or less, surrendered any hope for a communicable understanding of why so and so actually transpired.

To be sure, this does not mean that we cannot wildly speculate any number of possibilities for the odd event, but rather that we "test" those speculations in the empirical world. If we fail to do this, and this seems to be habitual among various New Age practices and beliefs, we are then left open to an almost infinite array of competing stories that rely more on faith than reason.

In light of this context, I personally don't see anything whatsoever in the Edgar Cayce readings which suggests that something truly psychic or supra mundane is happening. But this does not mean that I think that Edgar Cayce is a fraud or consciously trying to deceive his audience. To the contrary, I think Cayce appears quite sincere, even if naive, about the origins of his gift. What is perhaps more important, however, is that Cayce has had a profound impact on many people from all walks of life who have found tremendous meaning and purpose in his readings. This ranges from those who were in direct contact with him to those who have only met Cayce through his writings.

Thus the Cayce phenomenon must be tackled in two different ways: 1) From a purely scientific framework. Do these experiences represent something genuinely paranormal? And 2) Regardless of their putative origins, what does Cayce "mean" to people? These are, I would suggest, distinct questions and should be handled as such. Otherwise, the tendency is to conflate the two and in the process obfuscating any clear answer that may be apparent to both.

In answering the first question, we must be careful not to be so cynical and so dogmatic that we do not fully investigate all of Cayce's readings. This is why Johnson's approach is so useful and why his book is a necessary prelude to any final indictments on the extra-sensory claims inherent in much of Cayce's predictions. It is one thing for me to think that there is nothing "spooky" going on in Cayce's life and work, but quite another for my opinion to be stretched into a final scientific pronouncement. I have a strong hunch, based upon my reading of Cayce, that there is nothing paranormal happening, but my hunch is merely that and cannot, and indeed should not, be construed as a final closure to the ongoing investigation of Cayce's ideas. Skepticism is an extremely valuable tool in the arsenal of the researcher, but it is a tool among many. Ironically, it is better to have more broad-minded investigators explore Cayce first than having either firm believers like a Jess Stearn or an I.C. Sharma, or hard core skeptics like a James Randi or a Paul Kurtz, try to lionize or debunk him. The reasons for this are simple: the researcher who is unsure of his/her position allows for more conflicting reports to come to the surface, whereas the researcher who is already certain—either pro or con—has a tendency to drown any report which doesn't buttress his/her views.

Thus the scientific investigation of Cayce's psychic abilities must not be prematurely "explained away" by skeptics who have not fully and thoroughly investigated his case. Yet, at the same time, the flowering of hagiography that appears to be growing year by year around Cayce should not hamper such an investigation. W. H. Church's novelizations of the Cayce readings are a prime example of what not to do with Cayce's legend. Such crossbreeding of fact and fiction may sell lots of books, but they substantially detract from an unbiased appraisement of the sleeping prophet.

The second question, where one asks how Cayce's life and work has provided meaning to thousands of individuals, is a more complex issue since it involves a wide range of human emotions. Unlike the first query, which I believe can have a final answer (psychic or sociological? paranormal or normal? prophet or folk psychologist?), the question of meaning is an open-ended investigation which by its very nature betrays any single or final response.

Edgar Cayce has become—whether he would have wished it or not—a religious figure. And as a religious figure he serves as a fulcrum for people's yearning to connect with the mystery of being, the sacredness of life, and the wonder of creation. Edgar Cayce has become a modern myth and because of that exalted status transcends the either/or question of genuineness that skeptics, like myself, want resolved. Even if Cayce's readings turn out to be nothing more than the misidentified projections of his own unconscious mind, the Cayce phenomenon will not disappear because for many followers it is not simply a question of psychic ability. It is, rather, a larger question of sacred meaning and purpose and how they have found both in their relationship with Edgar Cayce's life and work. For these advocates Cayce remains a numinous touchstone and not merely a litmus test for borderline science.

One of the more interesting, if controversial, features of Johnson's book is that he takes a two-track approach in evaluating Cayce's psychic readings. First, Johnson attempts to distinguish fact from fiction in Cayce's proclamations. This modus operandi, refreshingly different from most of the popular studies on Cayce which tend to fuse the two (see W.H. Church's conflations, for instance, in his book Many Happy Returns)*, allows Johnson to be both critical and sympathetic. Second, while freely admitting where Cayce has made mistakes, Johnson then looks for the possibility that there may be a deeper religious or spiritual truth buried within the narratives, even if they contain fictitious elements. This is a particularly powerful approach since Cayce's readings tended to be full of spiritual import. Indeed, it may well have been this spiritual aspect that attracted so many to become followers of Cayce's prolific readings. In this regard, Johnson's fascinating profile of Cayce's numerous religious influences (from Theosophy to Bhagat Singh Thind) illustrates that the readings arise from the current fashions of the time.**

What a skeptic may wish to find but doesn't is an airtight case for Cayce's paranormal ability. Yet this is not Johnson's fault, since he meticulously tries to substantiate Cayce's clairvoyance, as in the instance of a predicted passage in the Great Pyramid and the right paw of the Sphinx. In both instances, Cayce's information was shown to be inaccurate. Yet despite such disqualifications, Johnson rightly states that Cayce's material remains interesting as a cultural phenomenon despite its "scholarly implausibility." Thus, Edgar Cayce in Context is not so much a study of purported paranormal ability (the evidence being scant or non-existent), but rather an insightful look at how a genuinely sincere "prophet" can change the course of people's live even if his prophecies are not extra-sensory. In other words, Johnson has tapped into the spiritual heart of Cayce and shown him to be a man of deep psychological insights, if not paranormal ones.

Ironically, the finest endorsement of Edgar Cayce's genuineness comes from a most unlikely source: Baba Faqir Chand, the radical Radhasoami guru of Hoshiapur, India. Faqir Chand, who is well known for dismissing any miraculous claims made about his life and work, proudly displayed a letter from Cayce's foundation, the A.R.E. When asked directly about Cayce, Faqir and his successor, Dr. I.C. Sharma, argued that he was an authentic mystic. This is no slight praise since it arises from a lineage, which tended to dismiss almost all gurus as frauds. Hence, I think Edgar Cayce's readings will survive into the 21st century not so much as illustrations of psychic ability misread, but as psychological and spiritual documents which resonate with seekers interested in a larger synthesis of New Age thinking. In this light, Edgar Cayce emerges as one of the architects to the modern esoteric revival.

Finally, I think K. Paul Johnson represents a new breed of scholar sorely missing in the academic field. He combines an acute critical judgment with a deeply held spiritual empathy, a rare combination. I think it is for this reason that Johnson's first two books on Theosophy have altered the course of future scholarship in that area. I have no doubt that Edgar Cayce in Context will do the same.

NOTES

* I have been conversant with the Edgar Cayce phenomenon for quite some time. I was commissioned by FATE Magazine back in the 1980s to review W.H. Church's then recent book on Edgar Cayce and his readings entitled, *Many Happy Returns*. I found the text atrocious primarily because instead of solely relying on the Cayce readings, Church decided it would be helpful to augment them with his own fictionalized interjections, including made-up dialogues that are not found in Cayce's own writings. To add insult to injury, Church argued that his additions were not really fictional. It is hard enough to wonder whether Cayce's original readings are genuine or imaginary without adding another layer of fantasy to them. Perhaps I can be forgiven for ending my scathing review of Church's book with the words, "The only happy return here for the reader is at the refund counter." Interestingly, a couple of years ago a television crew from Canada flew down to interview me for a featured program devoted to Edgar Cayce. Michael Shermer and I were chosen as the two skeptics to refute some of the more outrageous claims made about Edgar Cayce. I found myself in the unlikely position of partially defending Cayce, since I have long felt that he was not trying to perpetuate a fraud (like other charlatans I have known in the past, such as John-Roger Hinkins or Gary Olsen or Paul Twitchell or Sathya Sai Baba), but was rather genuinely sincere in his belief that he was just a medium for a higher power. But that doesn't mean that I think Edgar Cayce really had psychic powers. I don't. I just think we can all be deceived about the origins and causes of inner voices, visions, and the like.

** It is a little known factoid that Edgar Cayce was a personal friend of Dr. Bhagat Singh Thind, a spiritual teacher who taught a variant of shabd yoga practice. They were so close, in fact, that Cayce followed some of Thind's breathing and meditational practices and advised others to do the same. For further information about Thind and his interesting life and work (including allegations of plagiarism and genealogical dissociation) see Dr. Andrea Diem's *The Guru in America* (MSAC Philosophy Group, 2008).

POSTSCRIPT: A KIERKEGAARD CLARIFICATION

This is why I think Ramana Maharshi's state specific philosophy is an insightful and instructive one.

Elliot Benjamin's ends his essay explaining that he was somewhat confused with my claim that when "we are deeply depressed or extremely ill, we don't worry as much (if at all) about death or non-existence." He mentions Søren Kierkegaard and one of his more famous titles, The Sickness Unto Death, as an indicative counter ballast to my over arching claim. Yet, a close reading of Kierkegaard's insightful tome shows that he is focusing on a spiritual death, not necessarily a physical one.

My claim about severe depression or extreme illness was that in such situations, the idea of non-existence or the end of such suffering (by death or the cessation of consciousness) doesn't concern us as much, since what we desire in those moments is not "more" or "fuller" awareness, but less, much less.

Perhaps I can make my point clearer by using sleepiness as my example. When we are fully awake and happy and things are going well, we like the space we are in and we don't want it to end. Because of that, sometimes a fear arises that something will terminate it (the end of a beautiful vacation, the death of a lover, etc.). However, if we get really sleepy, our major concern focuses on when we are going to bed. Yes, we may still have fears and wants and desires, but their force lessens in direct proportion to how sleepy we are. Get sleepy enough, and philosophy and its urgencies disappear in a nocturnal flash.

Likewise, get sick enough (and here I can speak from firsthand experience, particularly on a few research trips to India where I got dangerously ill) and the great, unanswered questions of existence take a definite back seat to the desire for non-consciousness or deep sleep.

Another example that may help underline what I am driving at here comes from my youngest son, Kelly, who is (in the words of his current 2nd grade school teacher) an "old" soul and much more serious and much more philosophical than most adults I have met. I have noticed that whenever he is happiest he will come to me and remind me of a dream that has haunted him since he was 3 or 4 years old. In this dream, he realized that those he loved most—his mom and dad—were going to die one day and this made him unspeakably sad. However, I have also noticed that whenever he is very sleepy or quite under the weather (with a bad cold or fever), he tends never to mention the dream to me. Why? I think the answer, though it may not be universal, is a fairly obvious one: the fears we have are directly correlated to certain forms of awareness and their respective intensity. This is why I think Ramana Maharshi's state specific philosophy is an insightful and instructive one, especially as he raises the important point that only certain questions arise at certain times and even then with oscillating degrees of urgency. So when we probe deeper into the nature of self-reflective awareness (and its cyclical tendencies) we ask meta questions, such as "why do I ask why?" or "why do I fear death at twelve noon more than when I am really sleepy"?— simple and perhaps on the surface silly questions, but much more telling than we may wish at first to admit.

I also think Elliot Benjamin reveals something quite telling when he writes, "And along these lines, I must also say that I don't find it particularly comforting to know that I will eventually fall asleep and consequently all my unsatisfying ponderings about the lack of universal meaning in a materialistic universe will not be troubling me while I am sleeping—as long as I am not dreaming about it, which I apparently am safe from when I am in the dreamless deep sleep state. I mean, sure the extremely depressed person eventually gets some respite when he or she falls asleep, but this does not stop this person from committing suicide the next day."

Yes, it isn't appealing or satisfactory when we are fully aware and concerned (precisely the issue under discussion here and in our previous essay) to think right here and now that our questions will not be answered and that deep sleep will only give us a temporary respite. But that is exactly the argument: depending on our relative state of wakefulness (or alerted state of consciousness) we find some questions deeply pressing, whereas others concern us less, and vice versa. In other words, the questions that trouble us so much tell us more about our own state of being than about the universe to which we ask such questions. Immanuel Kant, of course, realized this centuries ago in his famous Critique of Pure Reason, even though he and others during this time period were naïve about neuroscience and its implications. In short, inspect the "I" that asks the question and see when and where and why it asks such at particular moments and not at other times.

This type of self introspection unravels the Gordian knot that seems to bind us to what has been rightly called the first human error: the confusion of neurology with ontology or the conflation of one's state of mind with the "real" state of the cosmos. We never absolutely know when we go to sleep if we are going to awake again. We assume such, of course, and over time we have become so accustomed to such repetitions that we have deluded ourselves about consciousness as a seamless and continual process when, in fact, it is anything but. Consciousness interruptus (or the vagaries of self awareness) show us moment to moment that we don't see the world as it really is. We cannot even see ourselves, much less grasp the totality of all that precedes and transcends us. Therefore, Socrates' most famous dictum, "Know Thyself" is a much more difficult injunction than we might at first surmise, since the very basis of the self and its varying energetic states blind us from comprehending them "as they are" but instead trick us into a perpetual hall of mirrors.

Wittgenstein touched upon this amaurotic tendency in human beings and the constrictive world of language in his philosophic lectures at Cambridge. The Zen Buddhist use of Koans (paradoxical questions or statements designed to usurp the rational mind) provides an even more illustrative understanding of why the questions we ask (and our unceasing attempts at answering them) are predicated upon an unenlightened predilection. We may think that such questions as "does a falling tree make a noise if there is no one there to hear it" are important or even answerable, but a textual analysis shows that the question itself rests upon unstated assumptions inherent in its construction. Unravel those assumptions first, and the question itself takes on a wholly different form. Likewise, unravel the uninspected assumptions we have about consciousness and the questions we have been asking from time immemorial will become transformed in the process.

TO CONCLUDE: A ZEN KOAN

A FAMOUS soldier came to the master Hakuin and asked: "Master, tell me: is there really a heaven and a hell?"

"Who are you?" asked Hakuin.

"I am a soldier of the great Emperor's personal guard."

"Nonsense!" said Hakuin. "What kind of emperor would have you around him? To me you look like a beggar!" At this, the soldier started to rattle his big sword in anger. "Oho!" said Hakuin. "So you have a sword! I'll wager it's much too dull to cut my head off!"

At this the soldier could not hold himself back. He drew his sword and threatened the master, who said: "Now you know half the answer! You are opening the gates of hell!"

The soldier drew back, sheathed his sword, and bowed. "Now you know the other half," said the master. "You have opencd the gates of heaven."

About the Authors

Andrea Diem-Lane is a Professor of Philosophy at Mt. San Antonio College. She received her Ph.D. and M.A. in Religious Studies from the University of California, Santa Barbara, where she did her doctoral studies under Professor Ninian Smart. Professor Diem received a B.A. in Psychology with an emphasis on Brain Research from the University of California, San Diego, where she did pioneering visual cortex research under the tutelage of Dr. V.S. Ramachandran. Dr. Diem is the author of several books including an interactive textbook on religion entitled *How Scholars Study the Sacred* and an interactive book on the famous Einstein-Bohr debate over the implications of quantum theory entitled *Spooky Physics*. Her most recent book is *Darwin's DNA: An Introduction to Evolutionary Philosophy*.

David Christopher Lane is a Professor of Philosophy at Mt. San Antonio College and an Adjunct Lecturer in Science and Religion at California State University, Long Beach. He received his Ph.D. in the Sociology of Knowledge from the University of California, San Diego, where he was also a recipient of a Regents Fellowship. He has taught previously at Warren College at UCSD, the University of London, and the California School of Professional Psychology. He has given invited lectures at various universities, including the London School of Economics. He is the author of a number of published books such as The *Making of a Spiritual Movement: The Untold Story of Paul Twitchell and Eckankar*; *The Radhasoami Tradition: A Critical History of Guru Succession*; *Exposing Cults: When the Skeptical Mind Confronts the Mystical*; and *The Unknowing Sage: The Life and Work of Baba Faqir Chand*, among others.

9 781565 432000